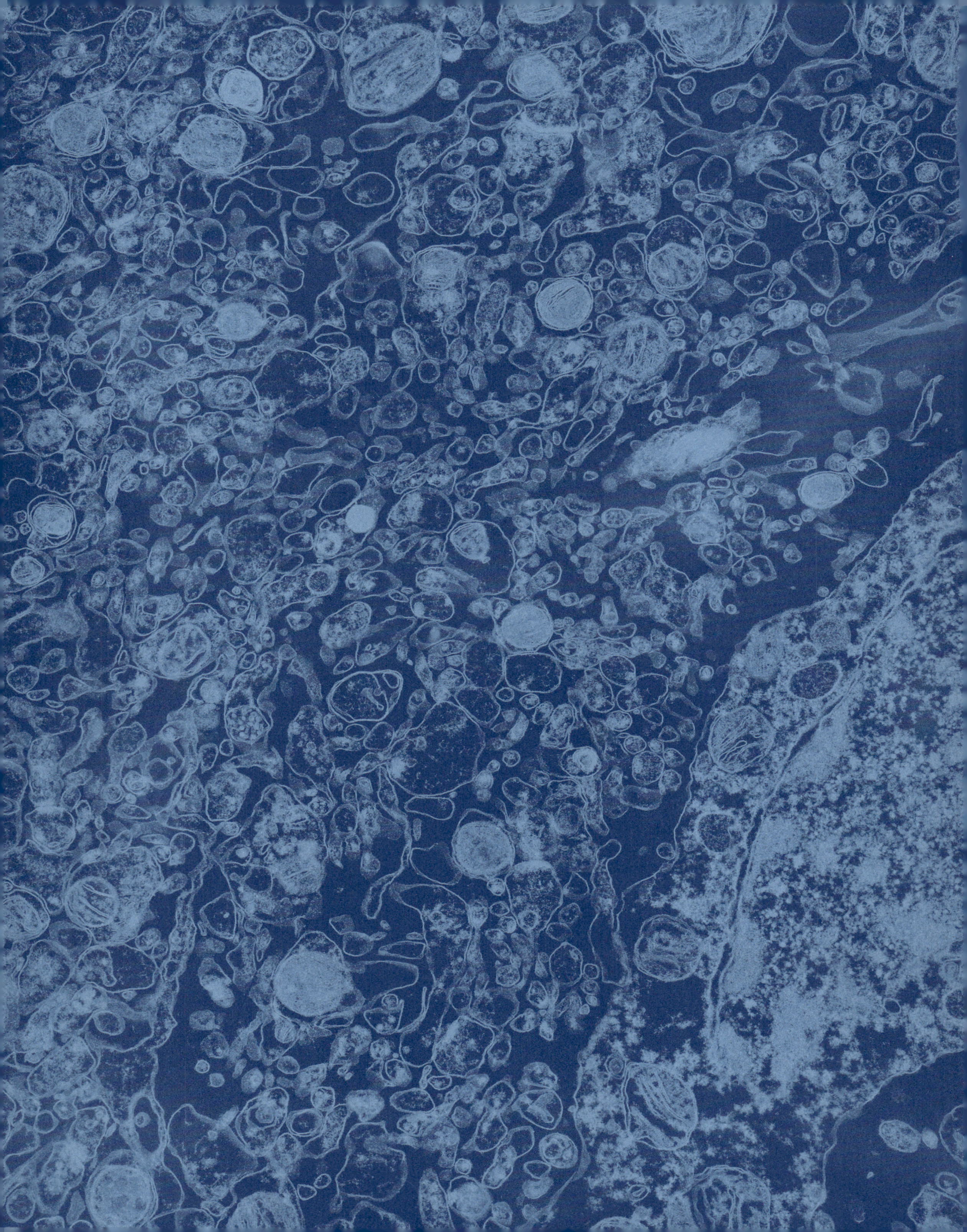

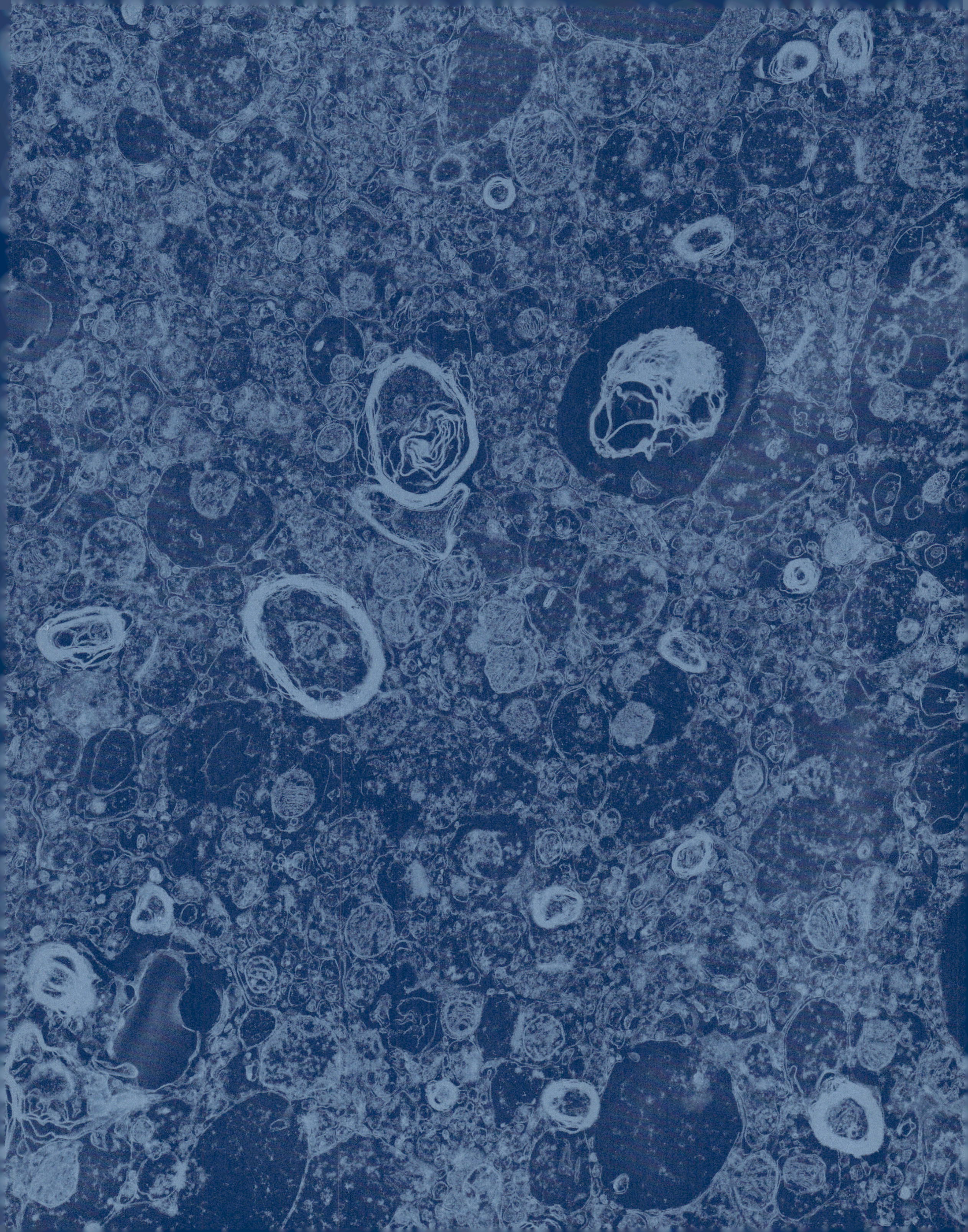

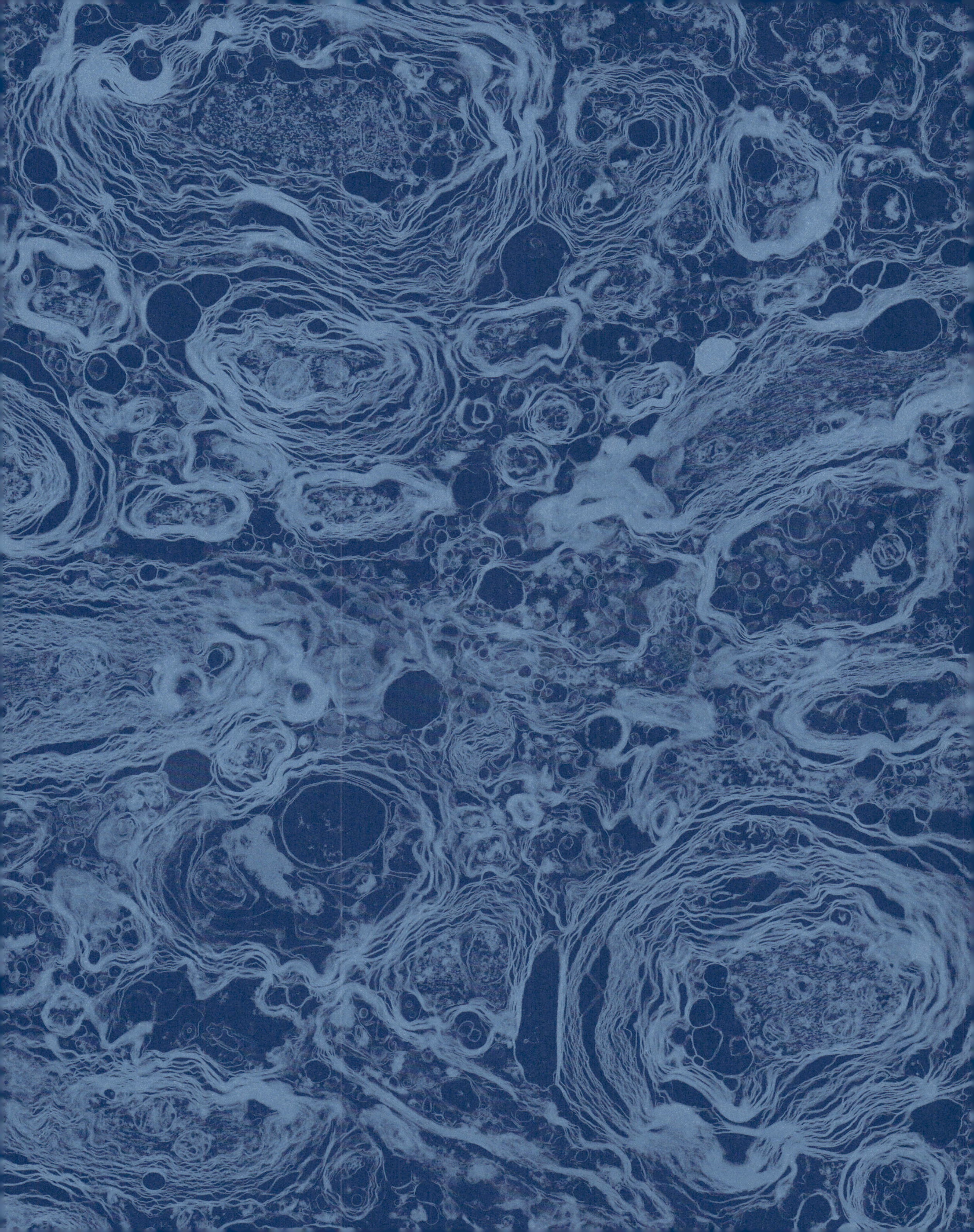

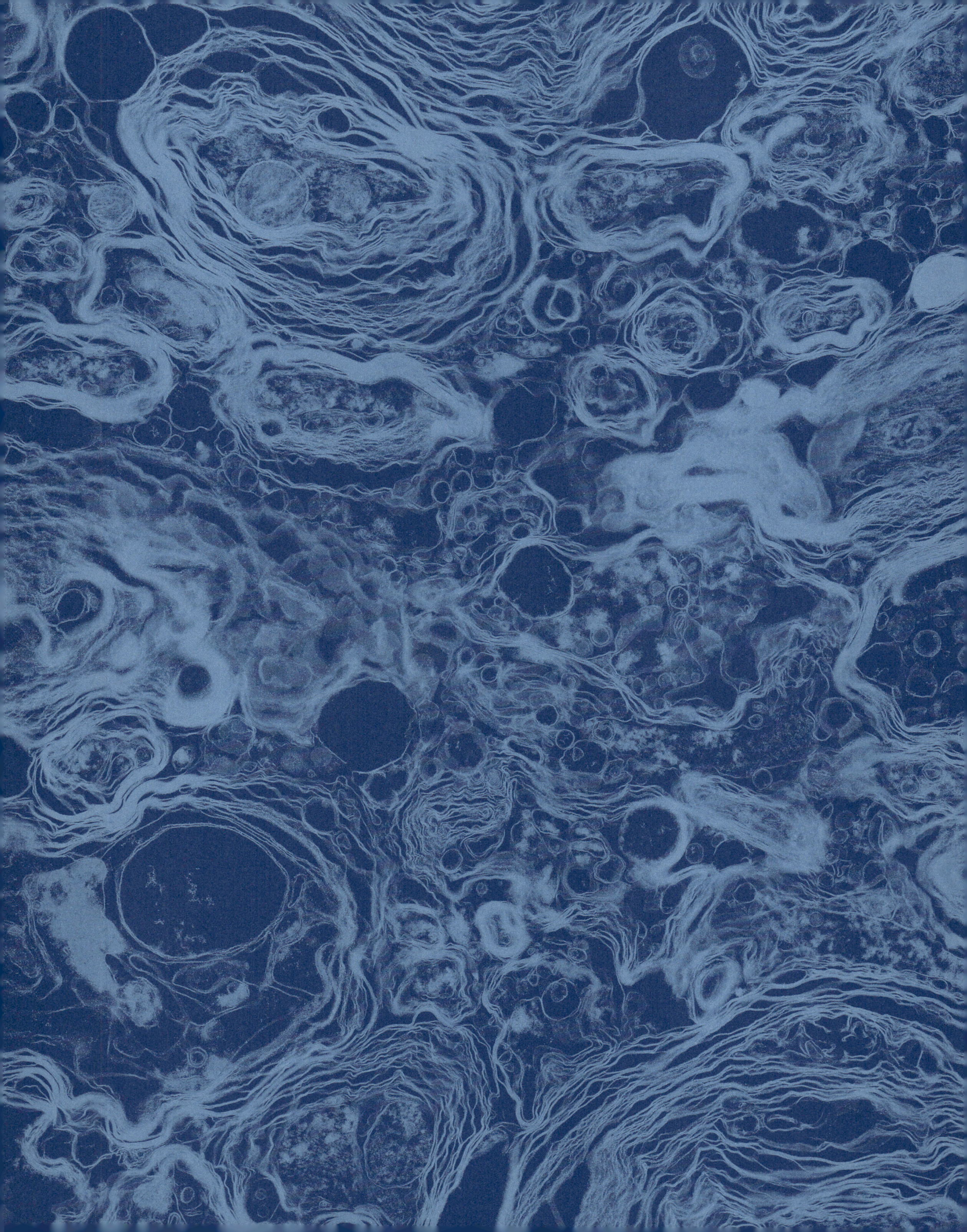

GOST

THE PROSPECT OF IMMORTALITY

MURRAY BALLARD

THE MARTINOT STORY

NUEIL-SUR-LAYON, PAYS DE LA LOIRE

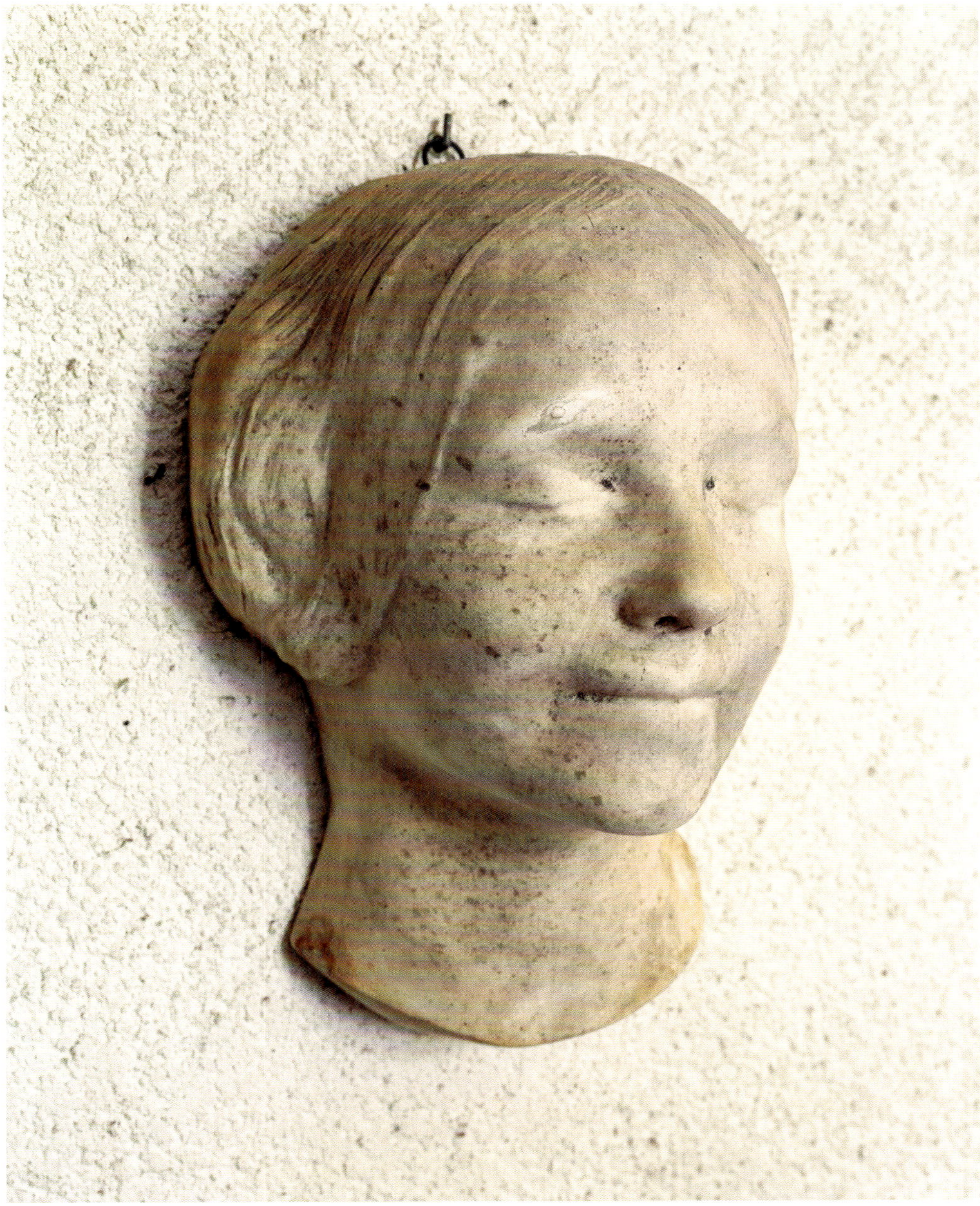

In 1970, Raymond Martinot, a French doctor who once taught medicine in Paris, purchased Le Chalet de Preuil on the outskirts of Nueil-sur-Layon, a small village in the Pays de la Loire region of France. His ambition was to create a working space to pursue his ongoing interest in cryobiology and life extension.

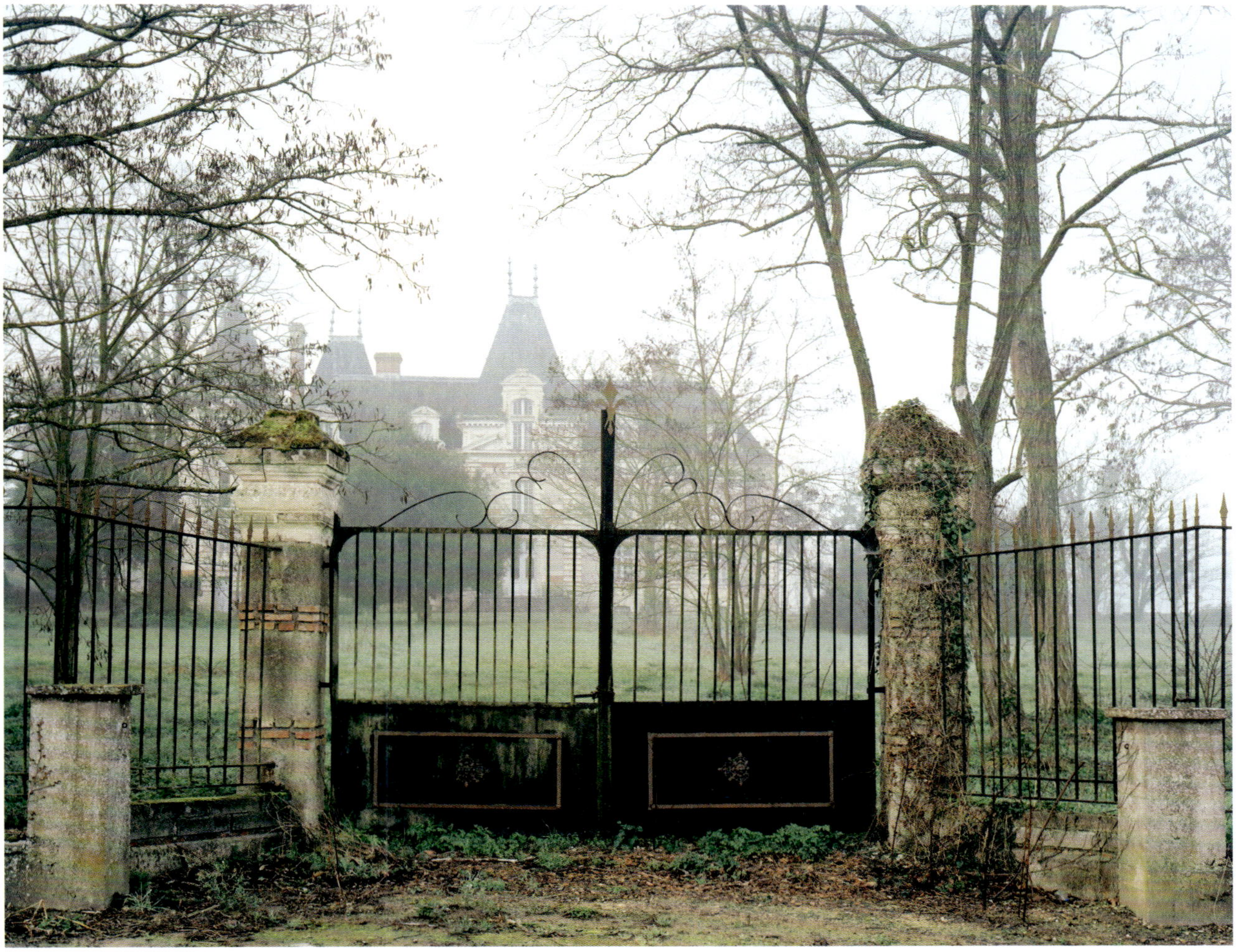

By 1974, he had designed and built a custom freezer to store his body after his death. He housed the freezer in vaulted cellars beneath the château. Martinot's theory was that whatever disease or illness caused his death could be cured by future medical technology. His estimate was that it could be possible by the year 2030.

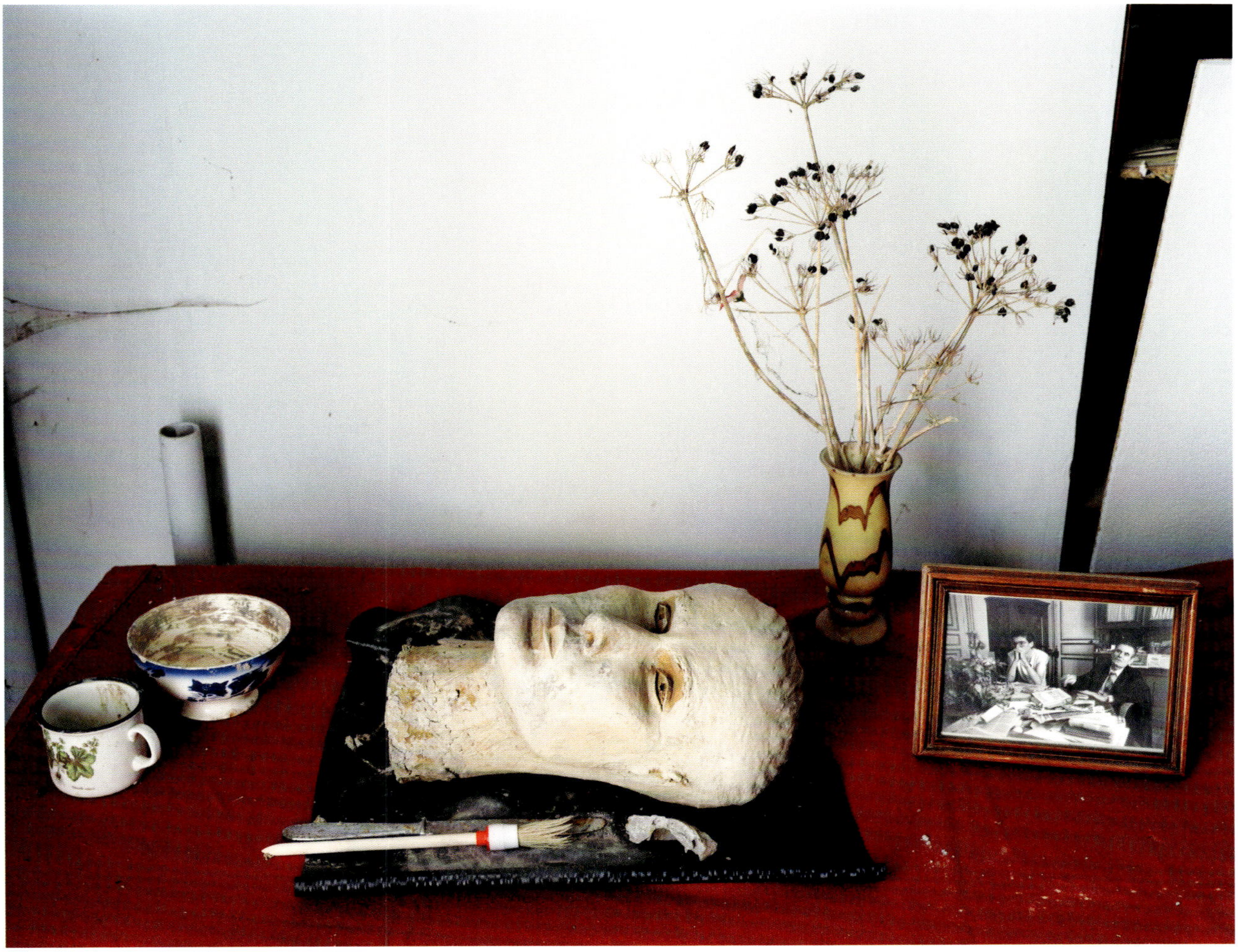

In February 1984, his wife, Monique, died of ovarian cancer. She was only 50 years old. Fighting through his grief, Martinot preserved Monique's body moments after she was declared dead, in the freezer he had constructed for himself. A lengthy legal battle over the ethicality of Martinot's actions followed, which lasted for almost two decades.

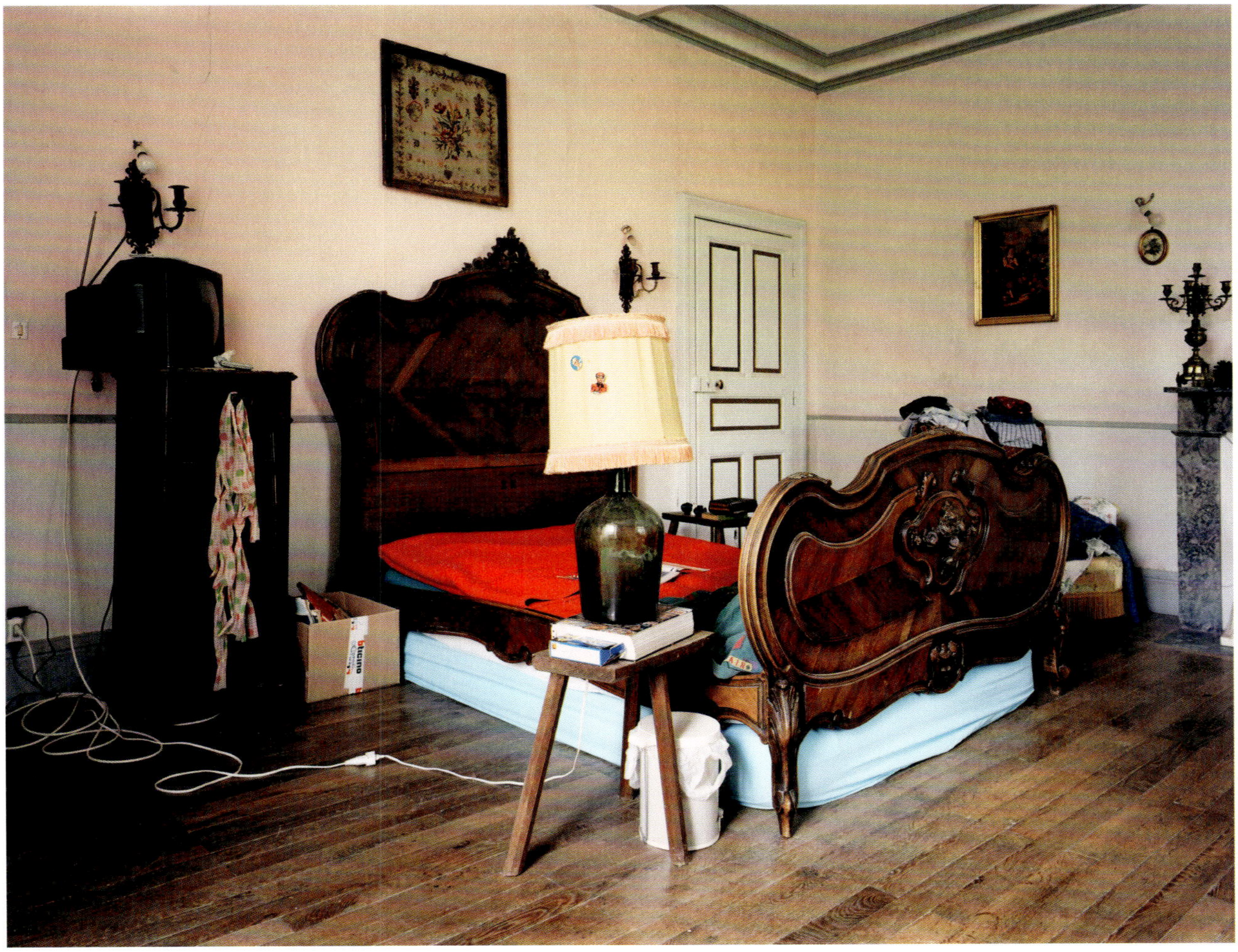

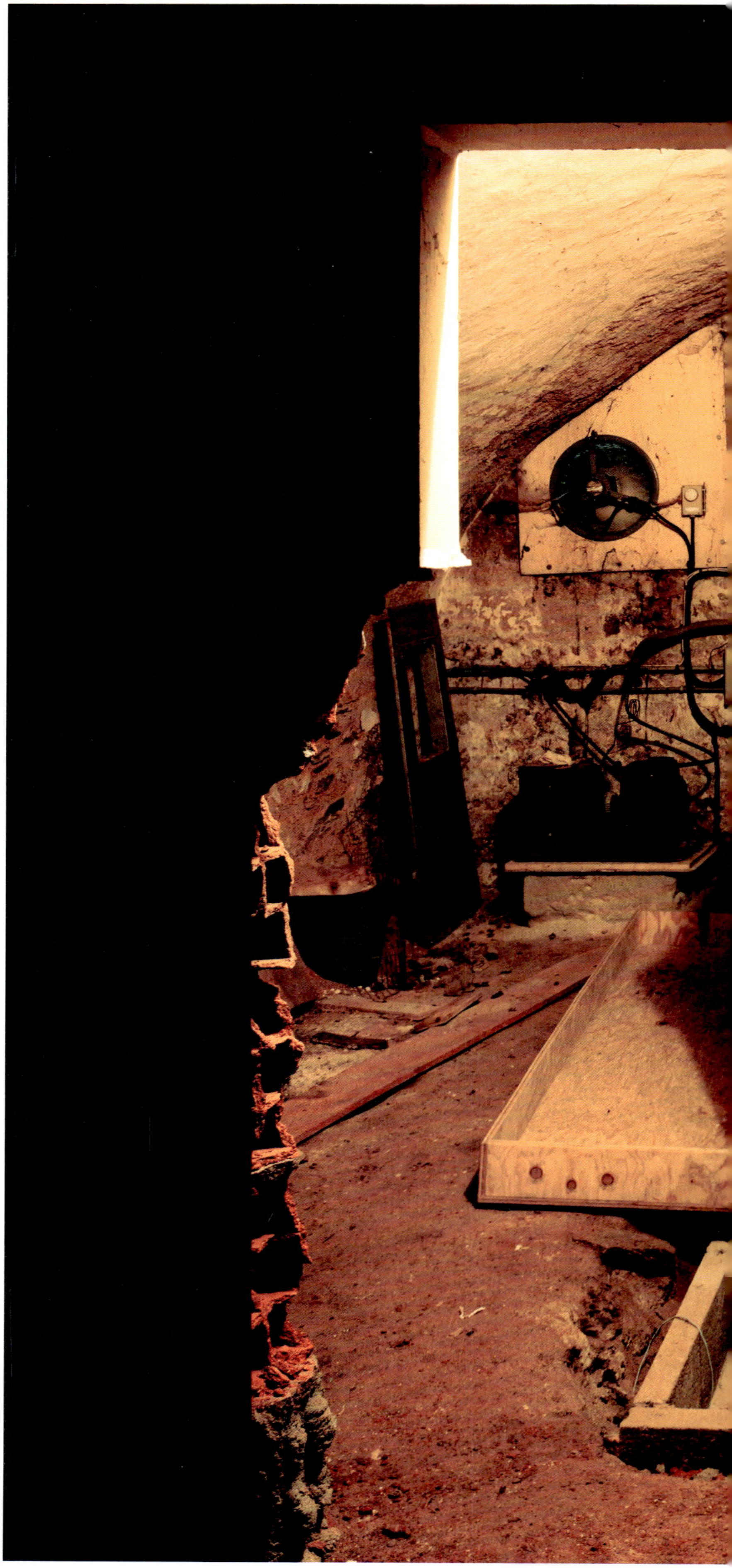

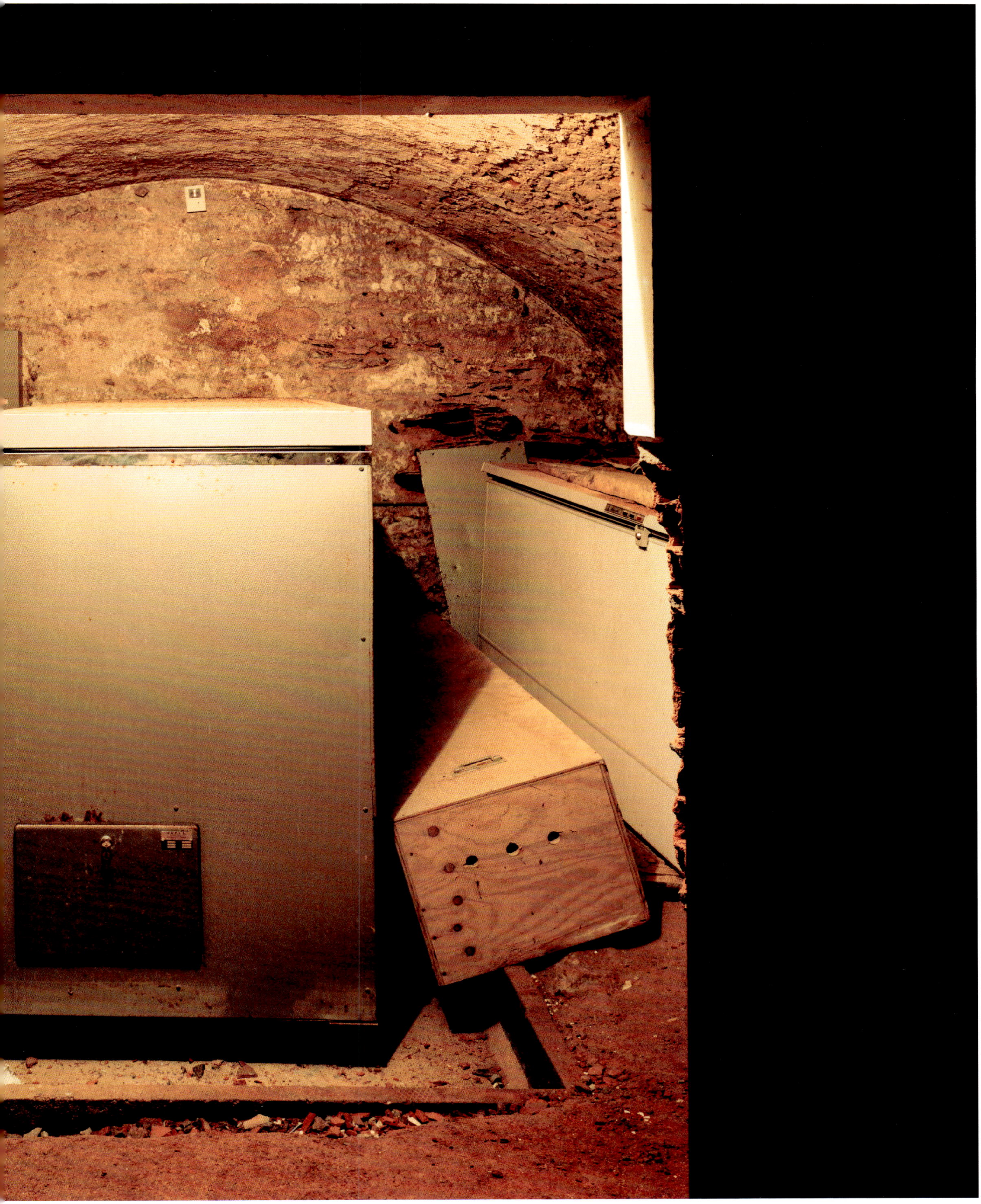

On 22 February 2002 Raymond Martinot died of a stroke, aged 84. His son, Rémy Martinot-Leroy, froze his body soon after. As instructed by his father, Rémy injected Raymond with cryo-protectants (human anti-freeze) and stored his body alongside that of his mother.

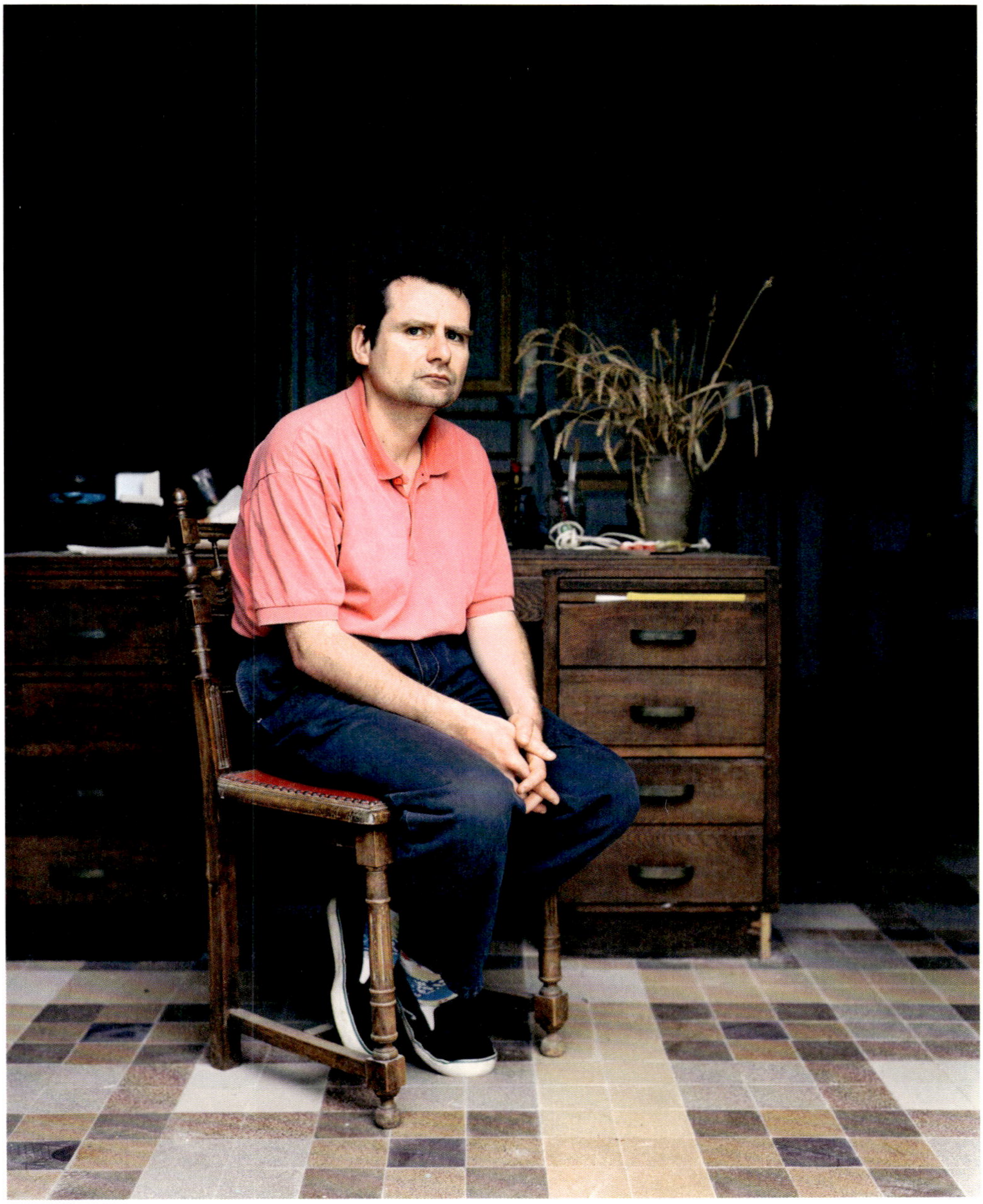

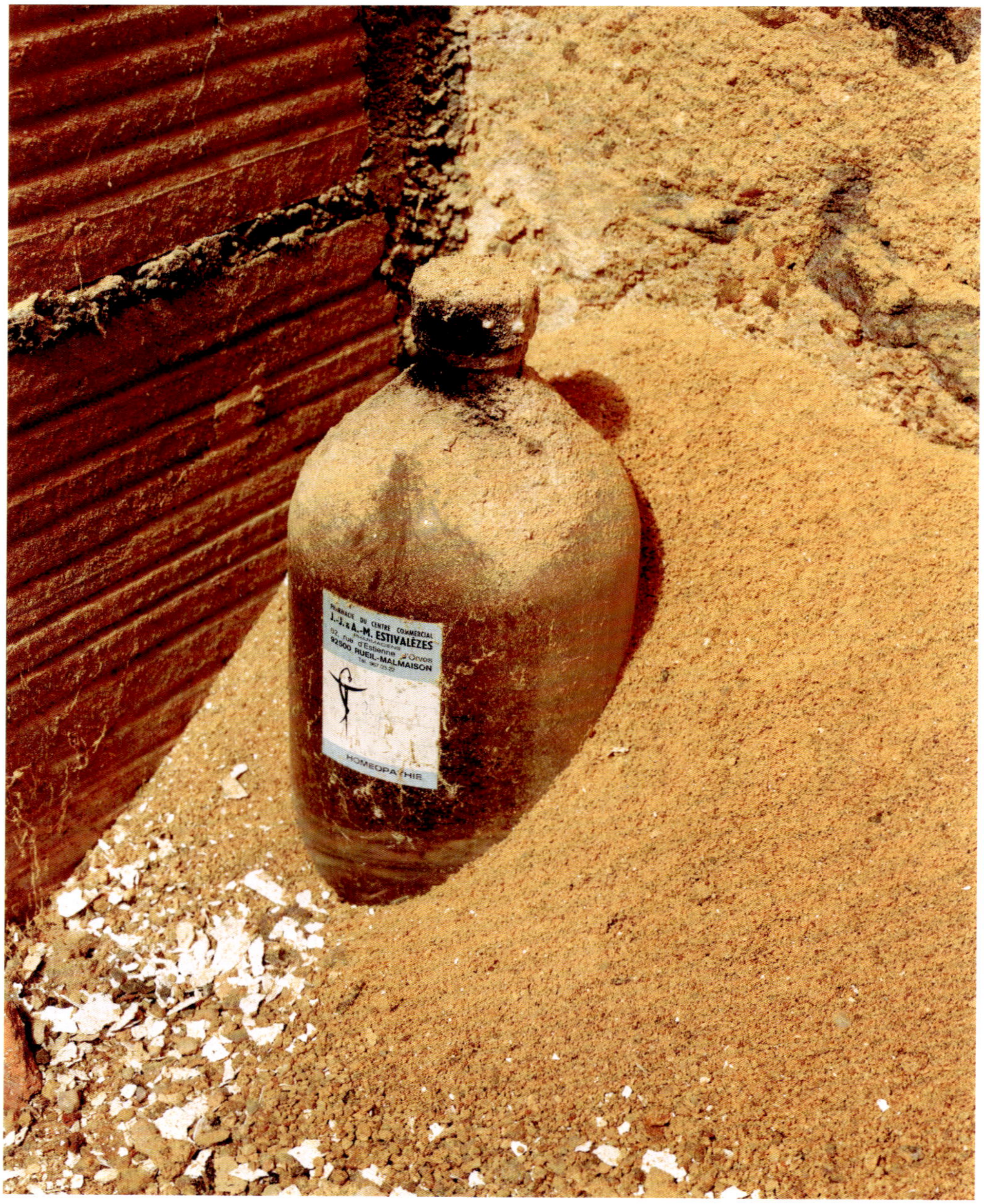
DU CENTRE COMMERCIAL
J.-J. & A.-M. ESTIVALÈZES
92500 RUEIL-MALMAISON
HOMEOPATHIE

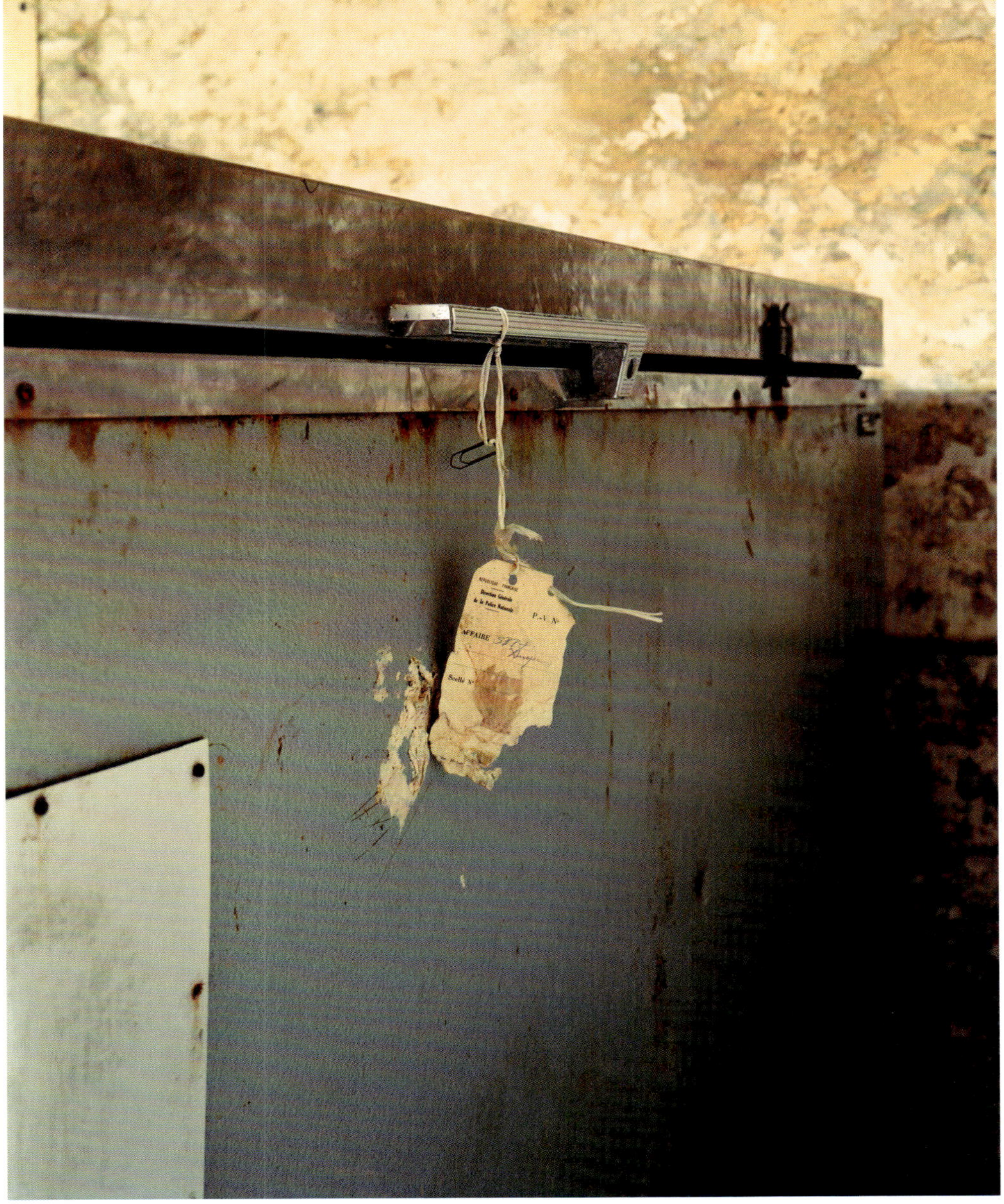
P.V. N°
AFFAIRE
Scelle N°

Rémy continued a legal battle in France's highest court to allow the bodies of his parents to remain frozen and stored in the cellars below the château. Despite his actions, the court proceedings were cut short when the custom-made freezer broke down and the bodies started to thaw-out. On 3 March 2006, Rémy had his parents' remains cremated, bringing to an end the 36-year experiment.

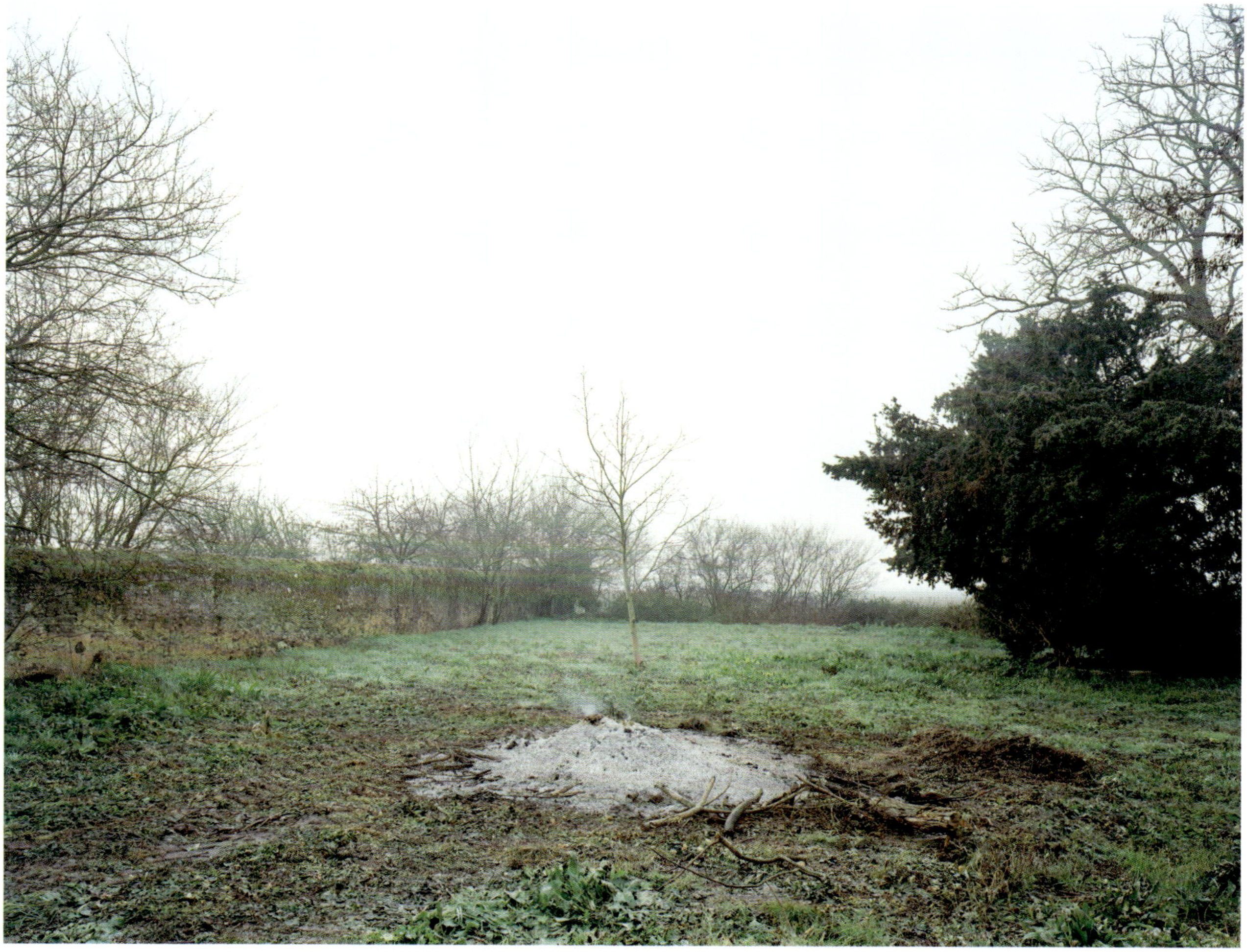

ALCOR LIFE EXTENSION FOUNDATION

SCOTTSDALE, ARIZONA

I have been in the field of emergency medicine for about 20 years. Those skills really led to my current position as the Medical Response Director for the Alcor Life Extension Foundation. Alcor, as it's commonly referred to, is a not-for-profit scientific research foundation, which is dedicated to the science of cryo-preserving bodies. We have about 1000 members around the world who have signed up and said 'I'm going to donate my body to this science project upon my death'.

We'll send a team to their bedside when the end is expected and, literally within 60 seconds of their death being pronounced, we're gonna start. We're gonna restore oxygenation. We're gonna restore circulation. We're gonna administer a number of post-mortem medications. We're gonna rapidly cool the body. The next level of death is legal death, which is when current medicine and science can no longer start the heart. It doesn't mean that every cell in the body has turned off — they just can't get the heart started again.

There is a third level of death that I think most cryonicists believe in and that's a 'theoretical' death. We really believe that the memory, identity and personality of the brain is similar to the hard drive of a computer. It doesn't go away just because you turn it off.

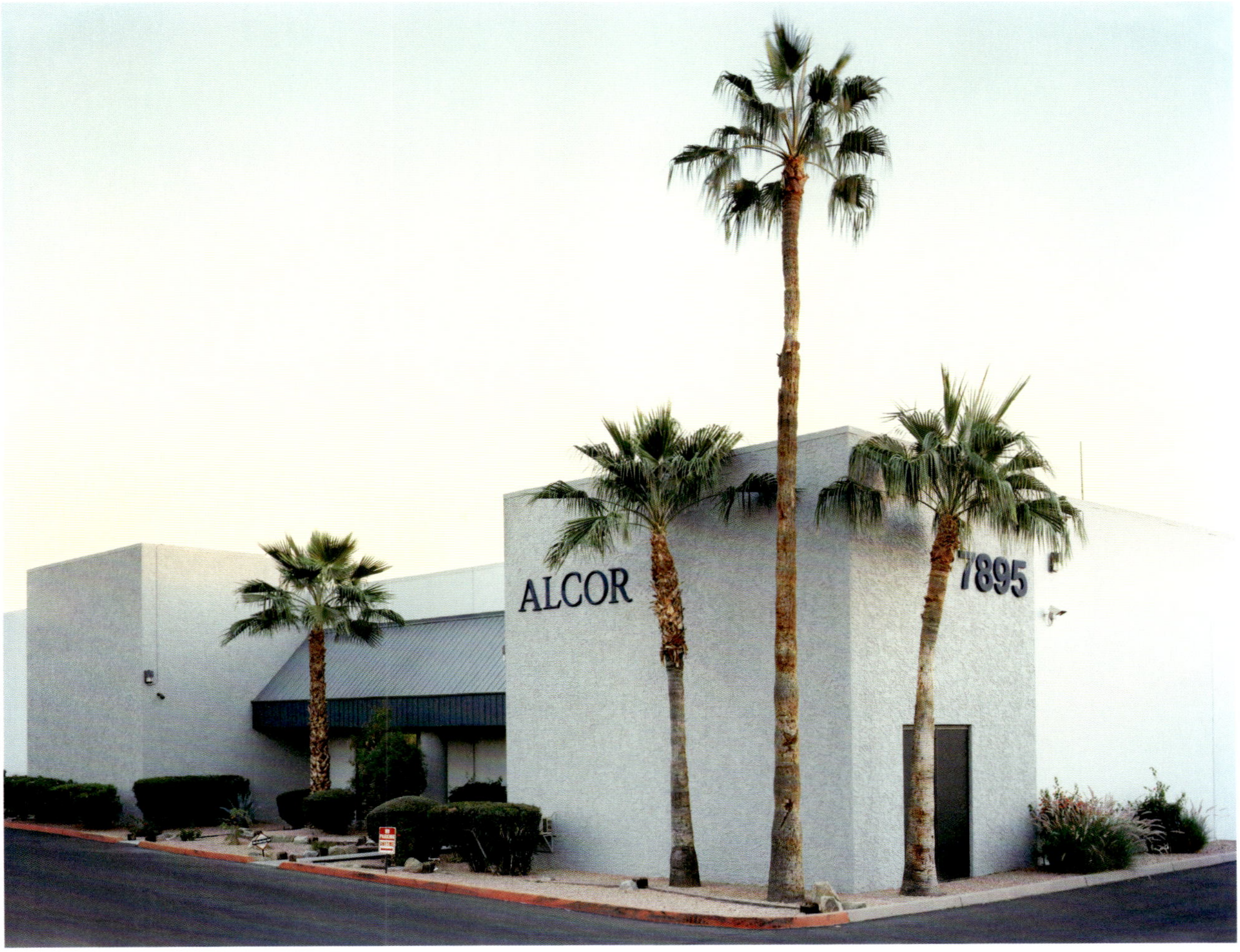
ALCOR
7895

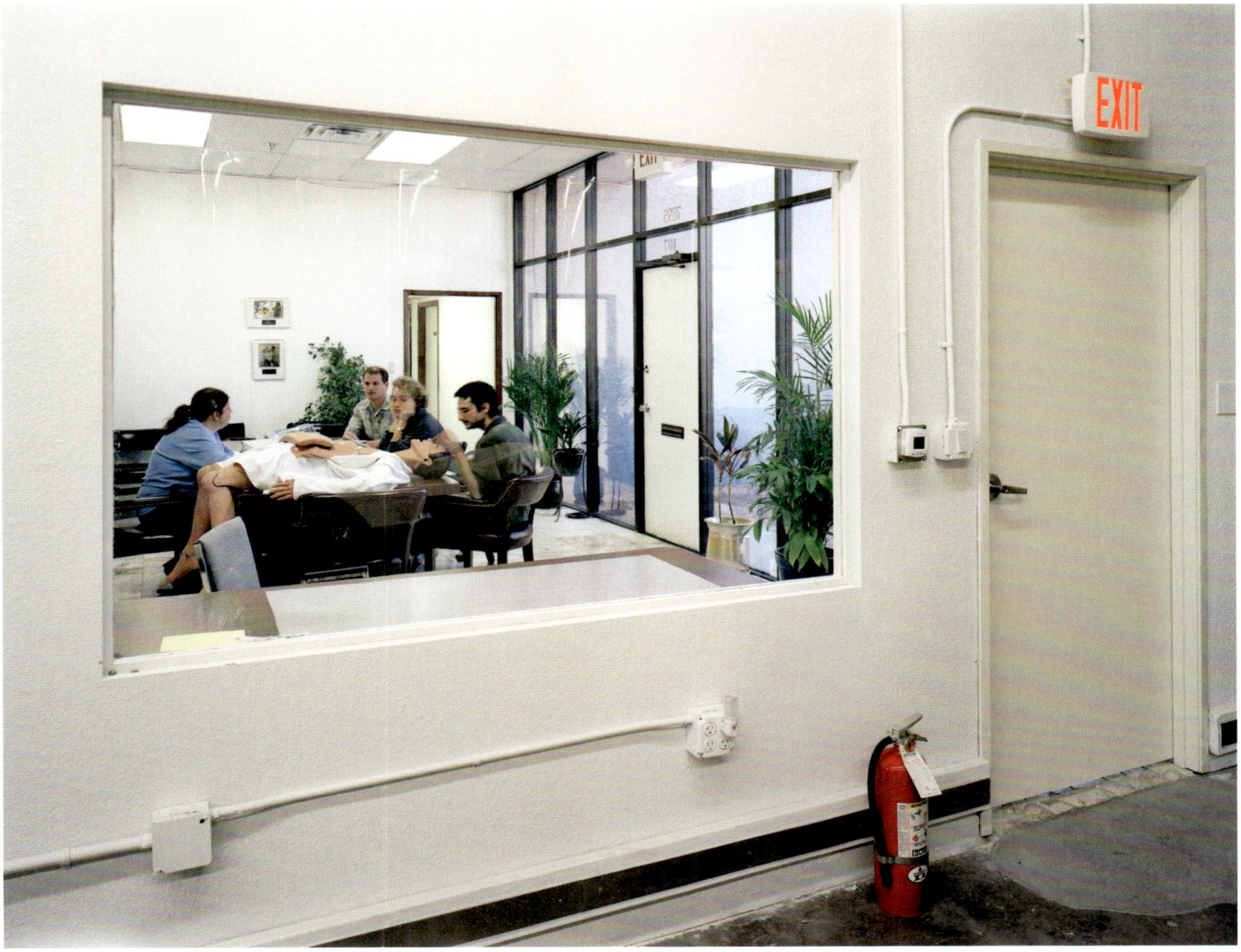
EXIT

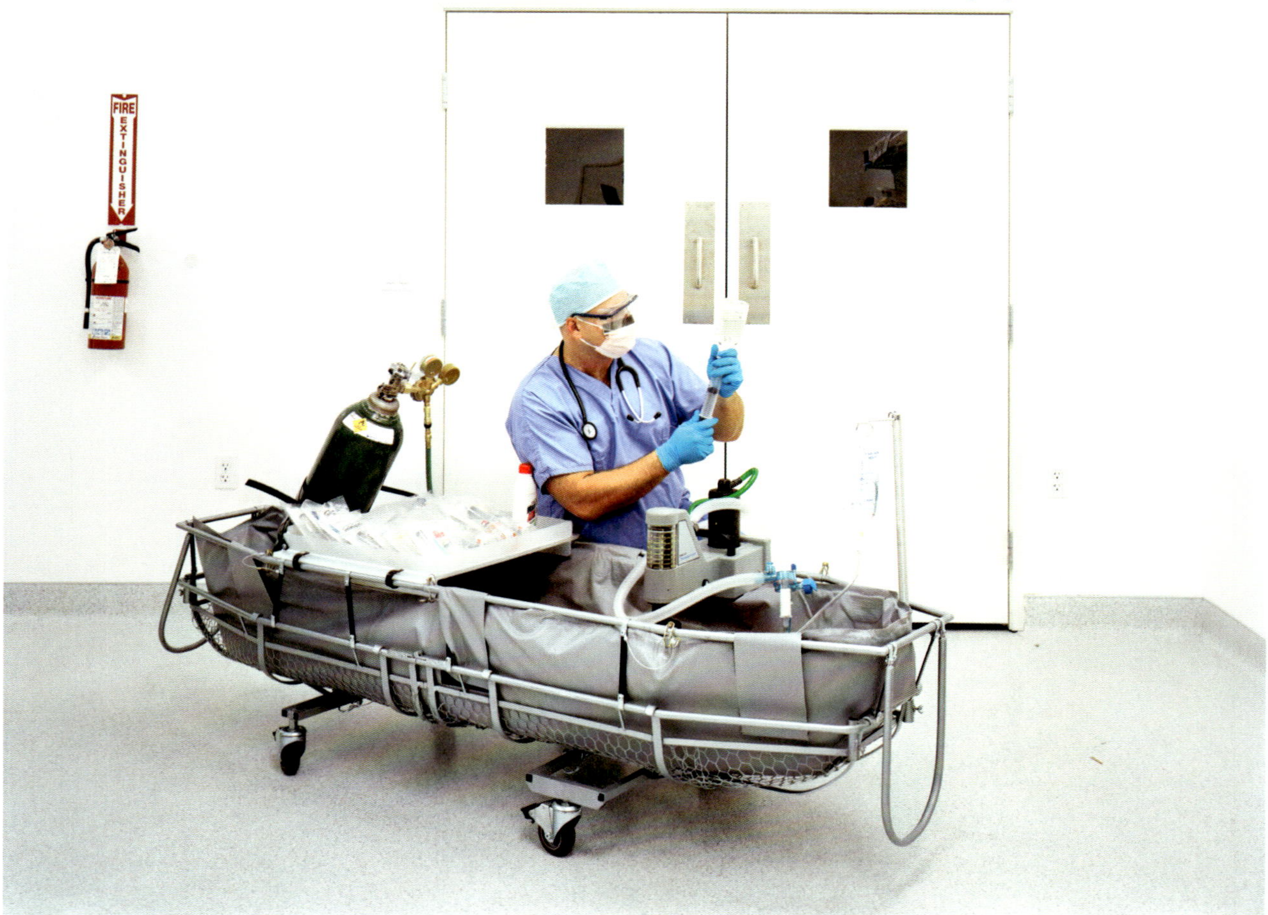
FIRE
EXTINGUISHER

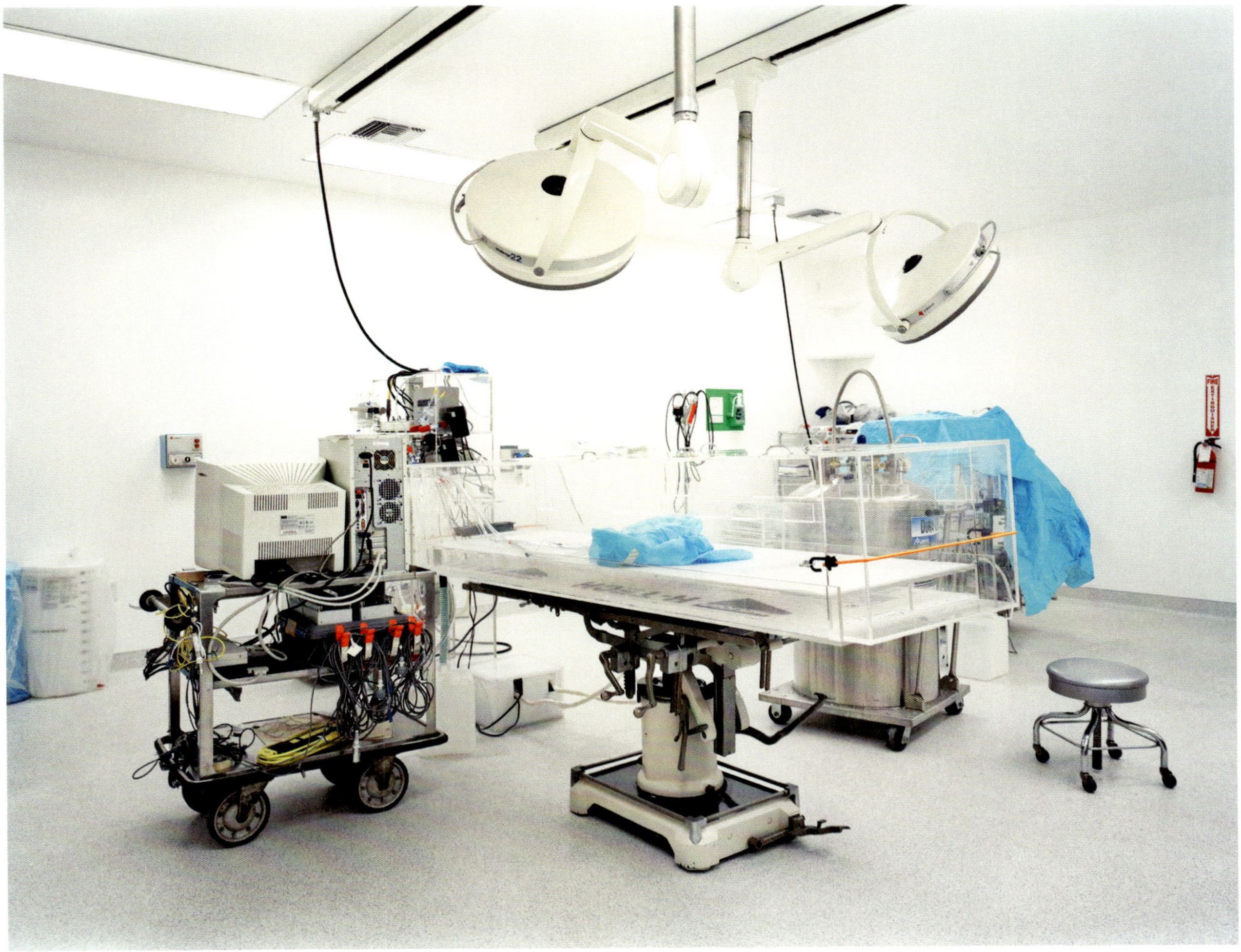

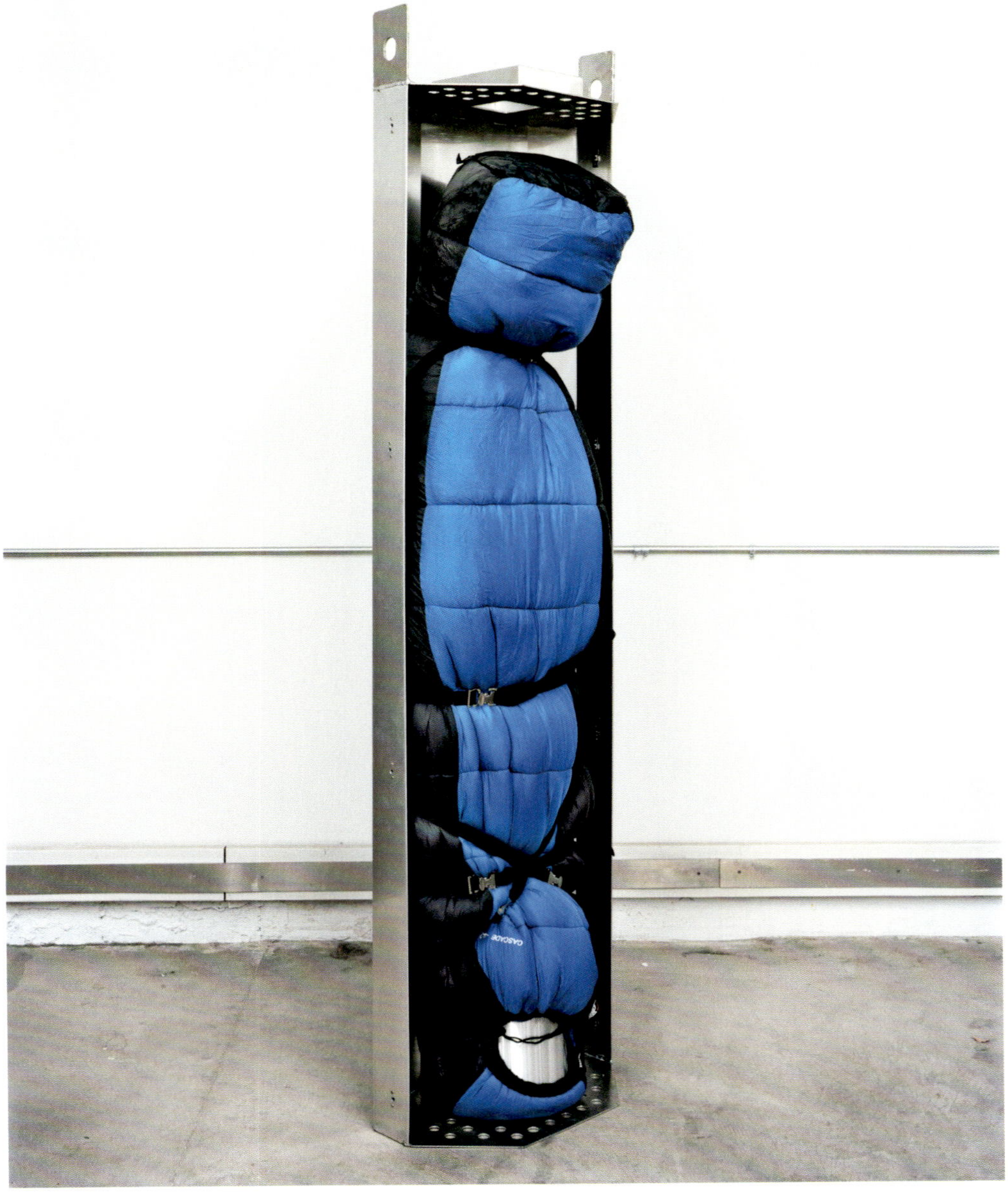

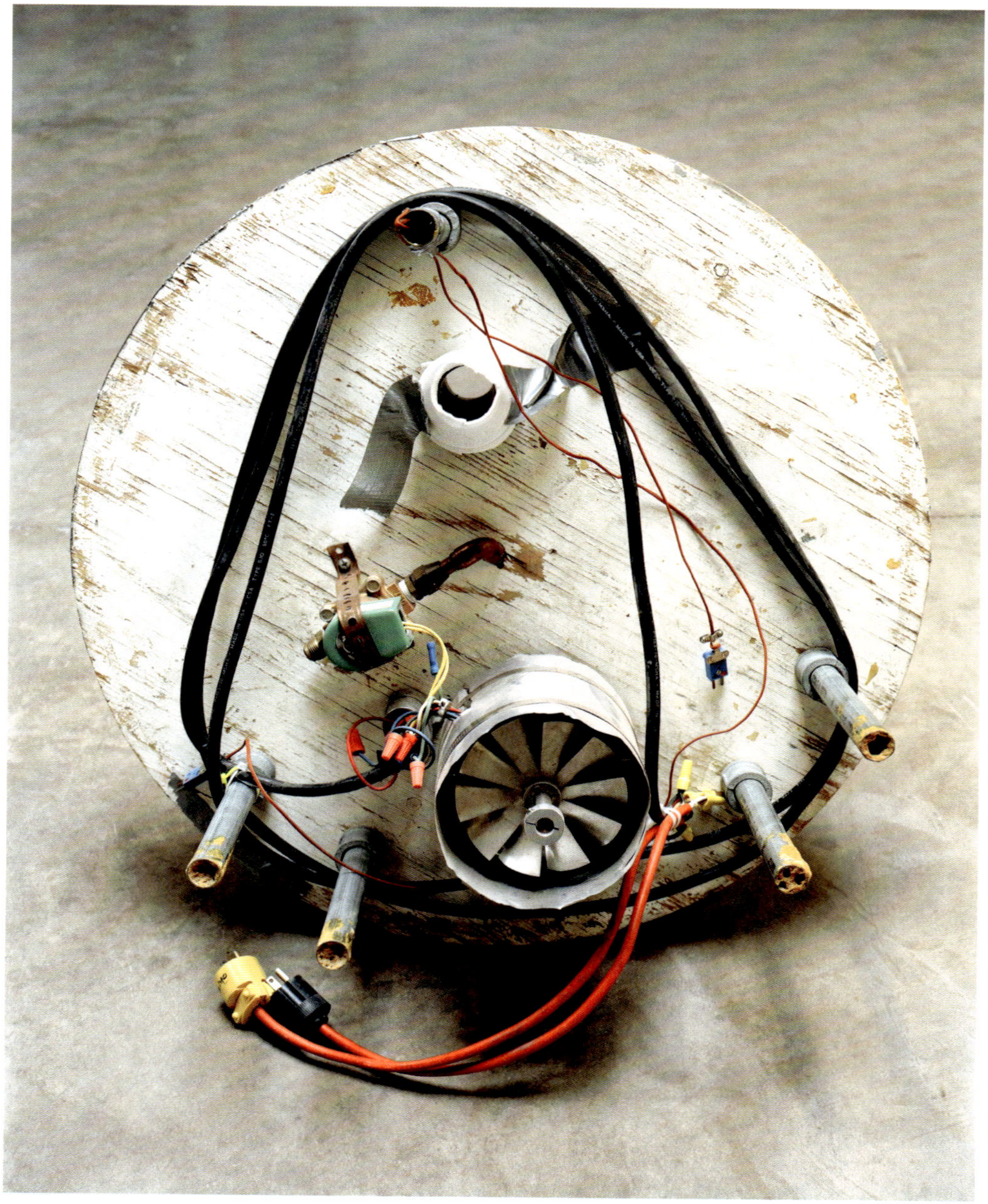

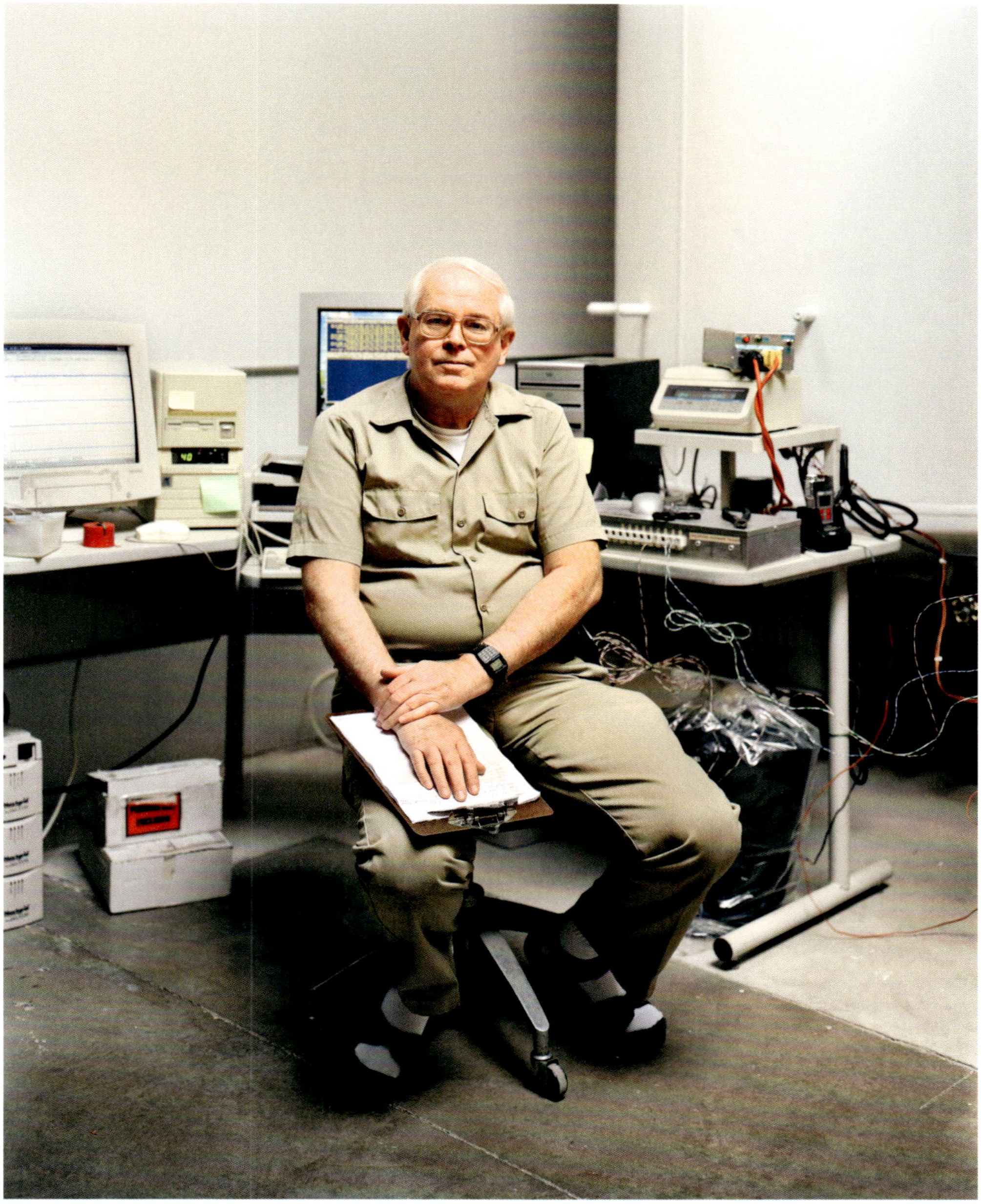

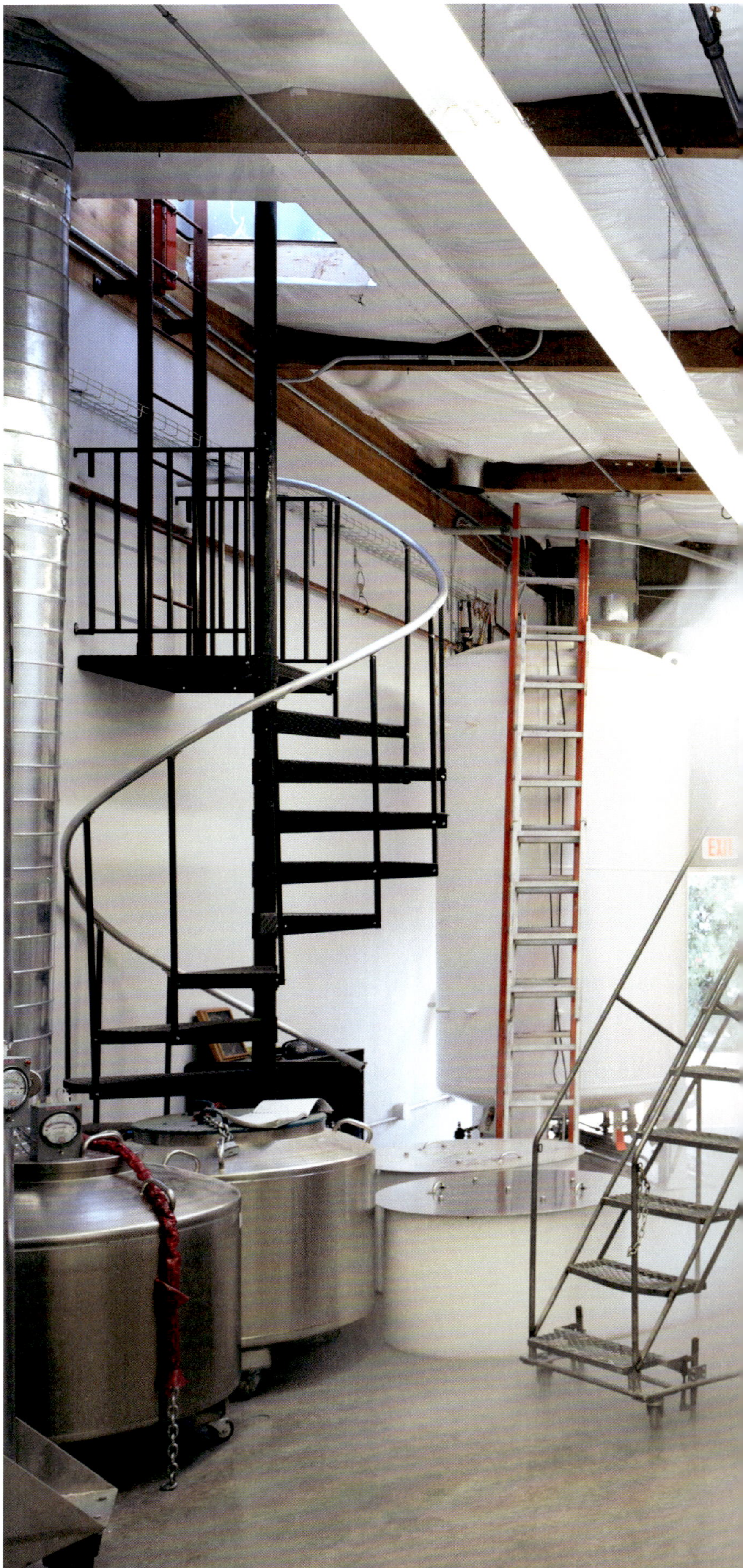

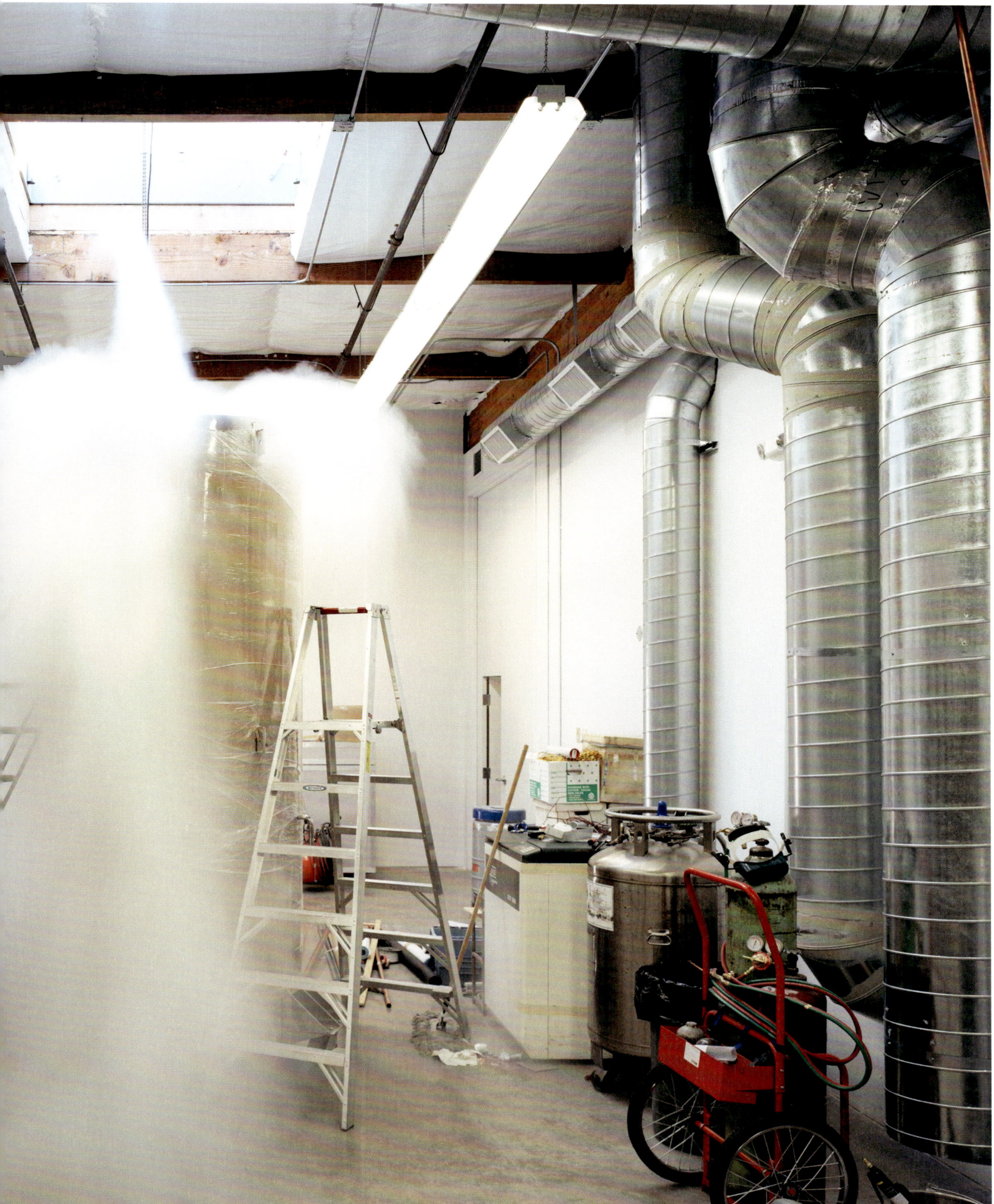

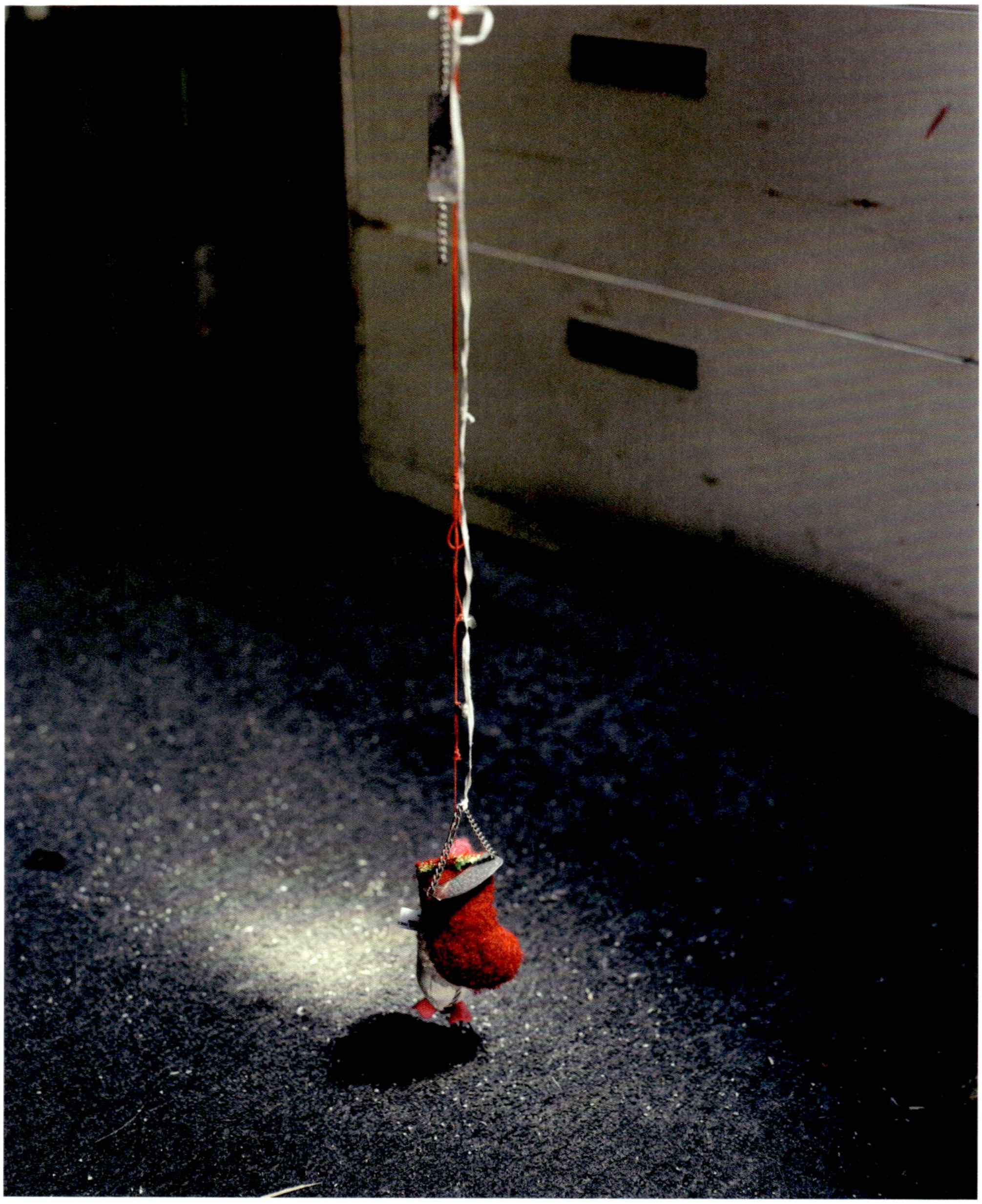

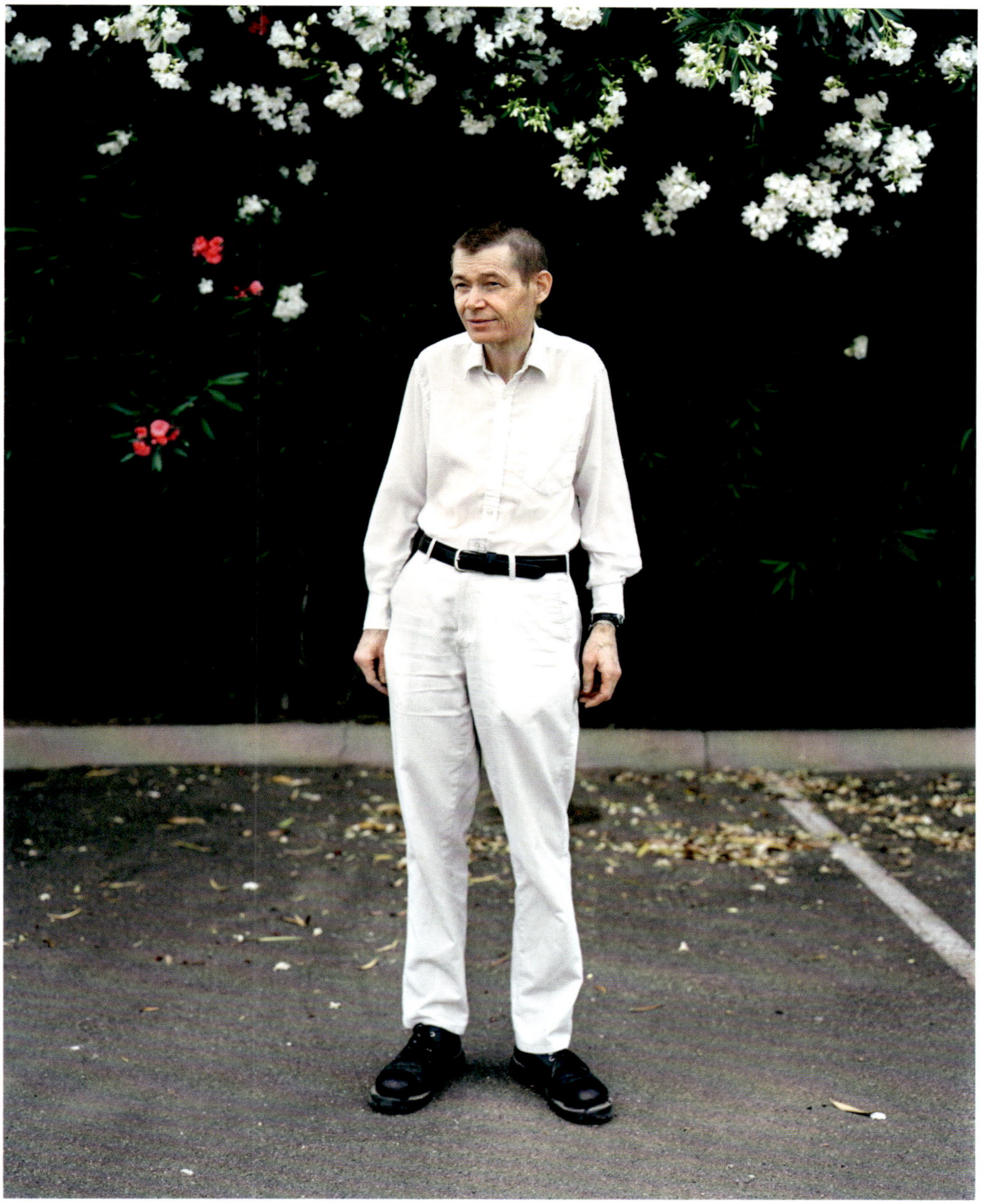

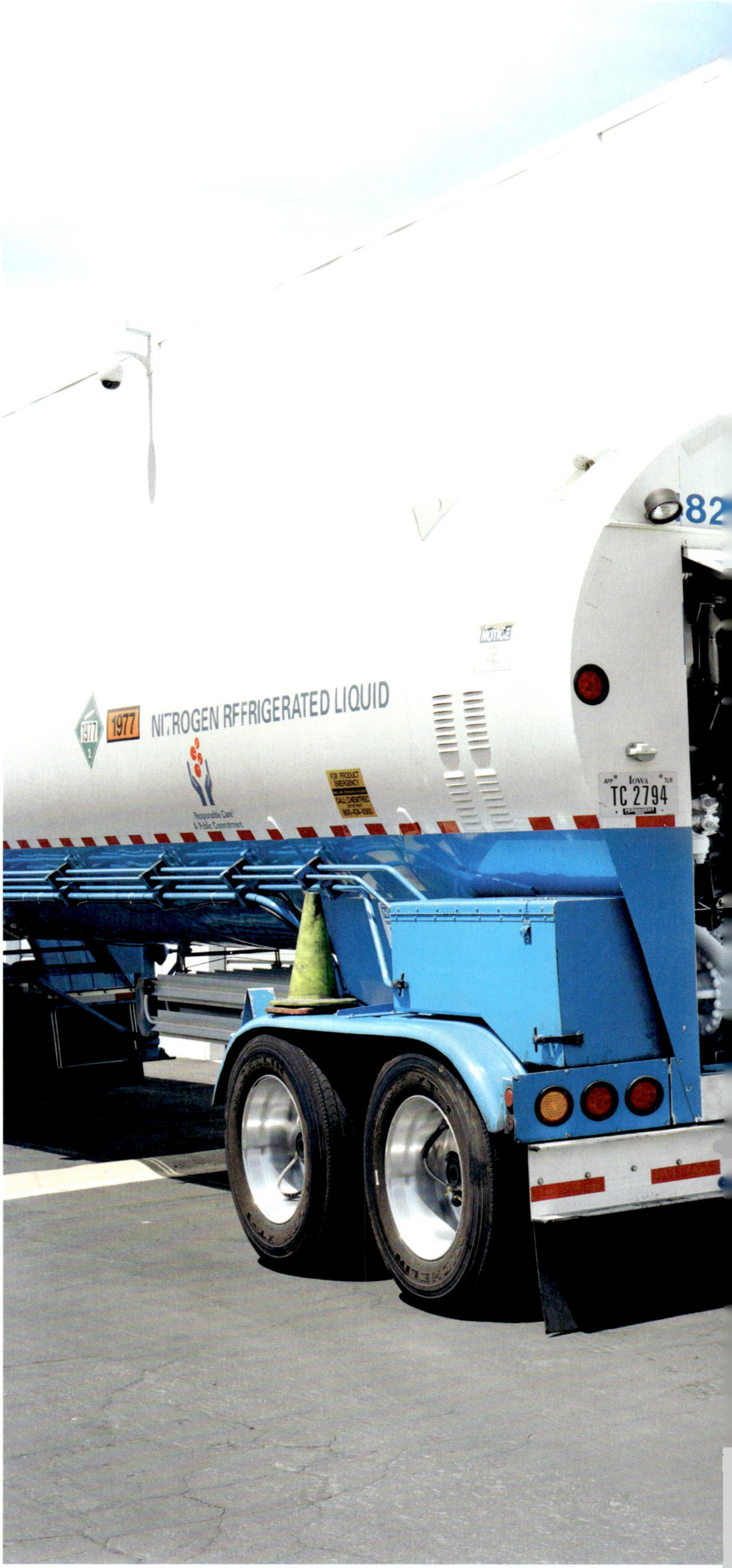
1977
NITROGEN REFRIGERATED LIQUID
TC 2794

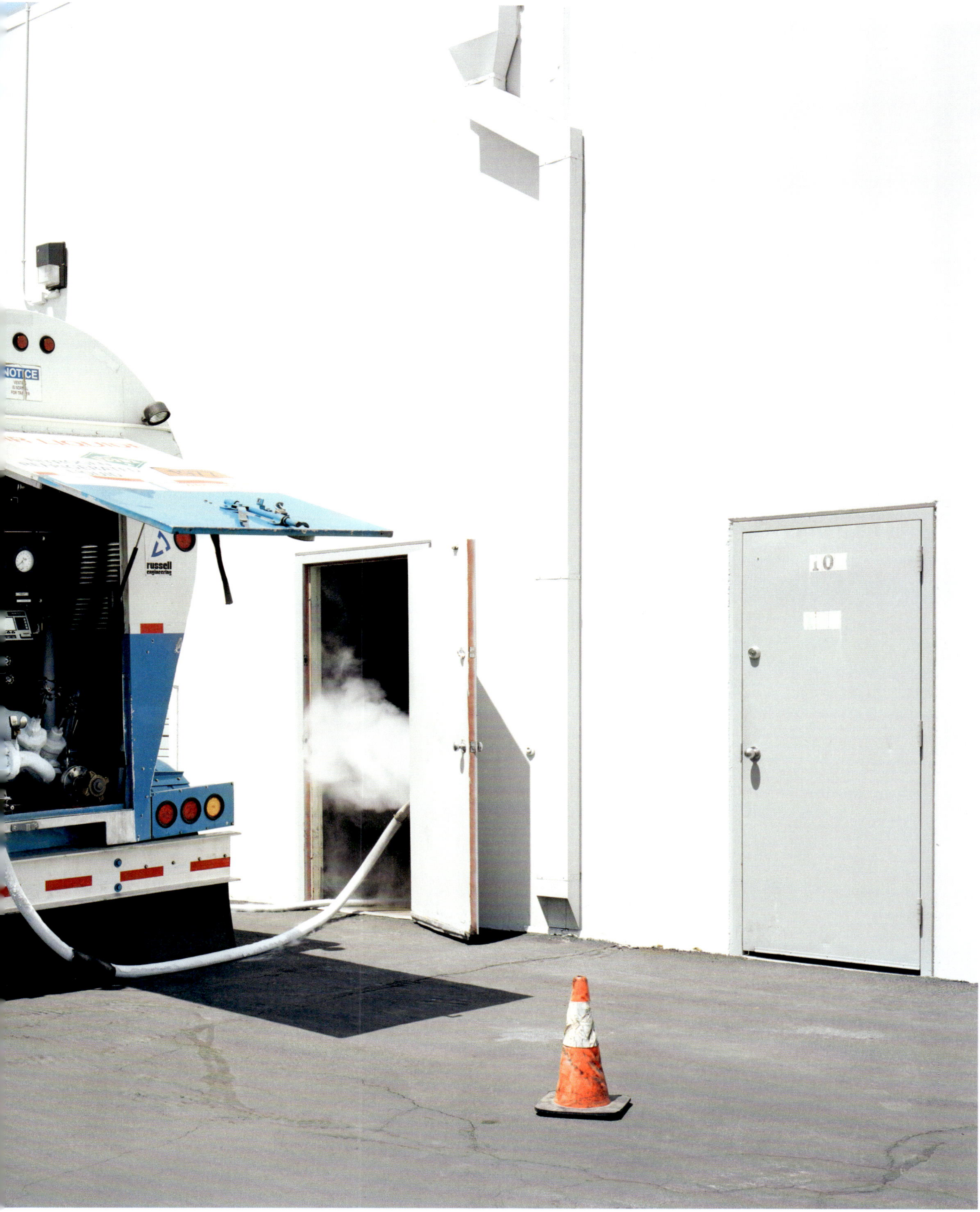
NOTICE
russell
10

CRYONICS UK

PEACEHAVEN / LONDON / SHEFFIELD

Cryonics UK is a local standby, stabilisation and transport team comprised of volunteers.

There are currently no cryonic storage centres in the UK and no trained staff to undertake the necessary posthumous procedures.

Members from around the country meet up and train themselves in the initial stages of a cryonic suspension.

These procedures include: surface cooling, cardiopulmonary support (CPS), blood washout and administering stabilisation medications.

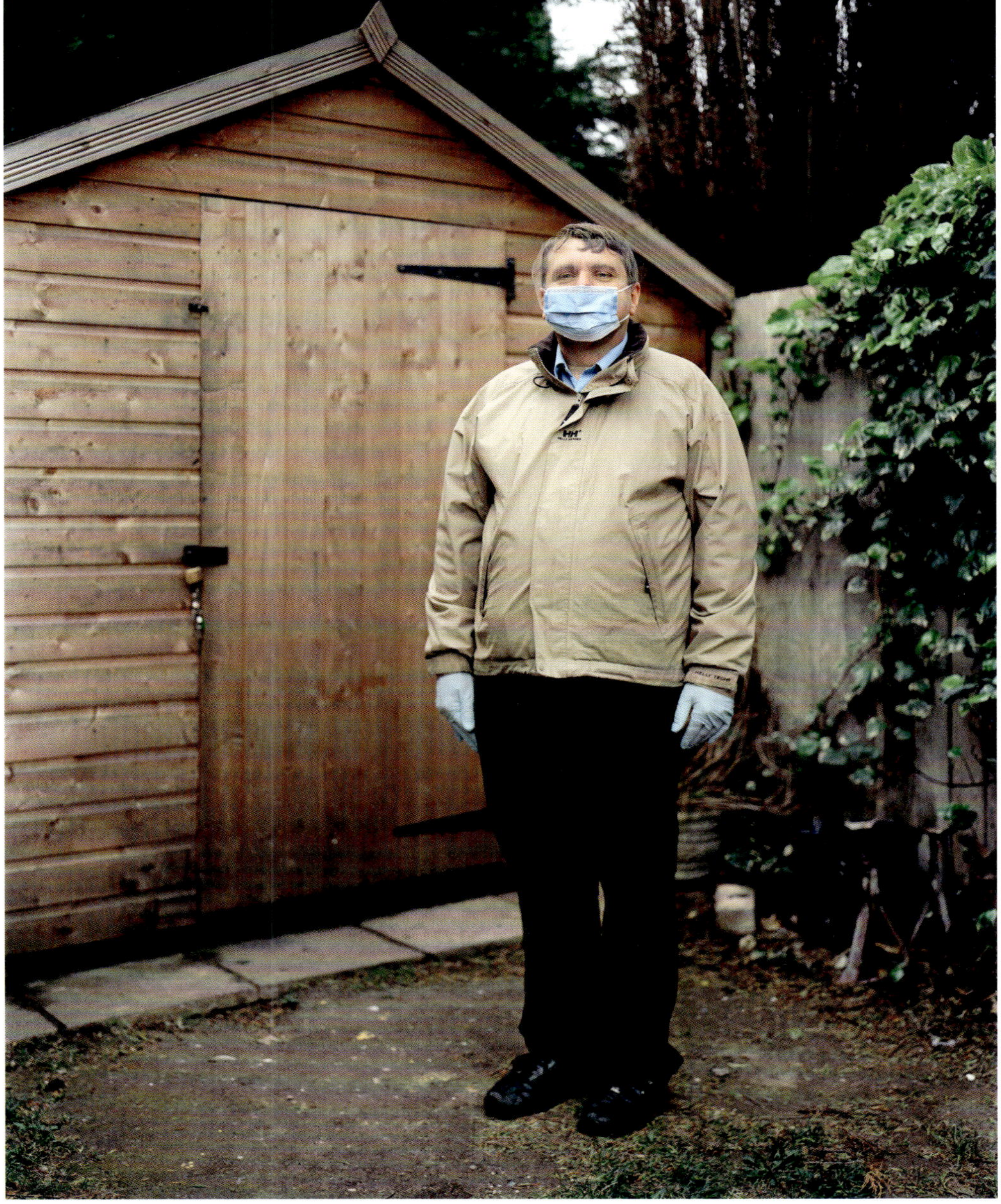

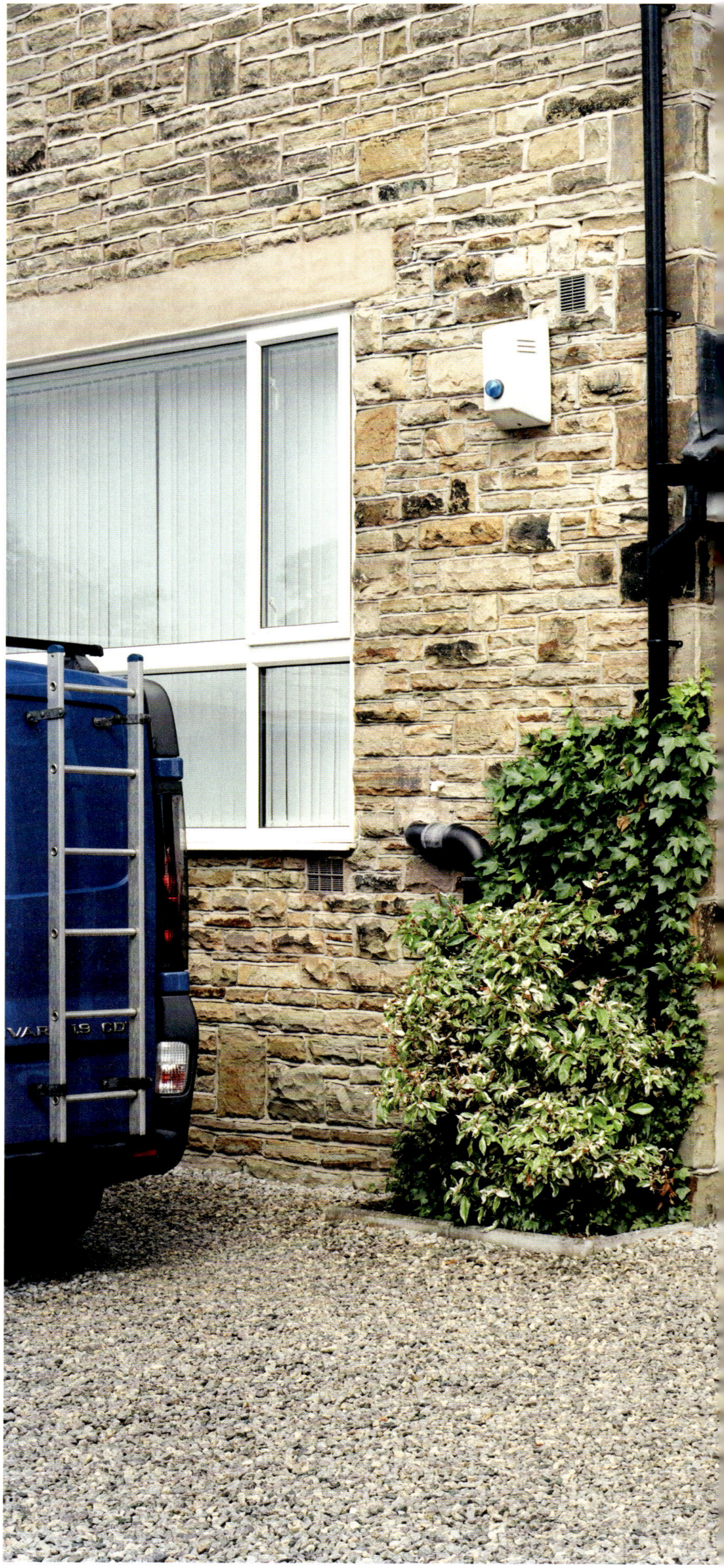

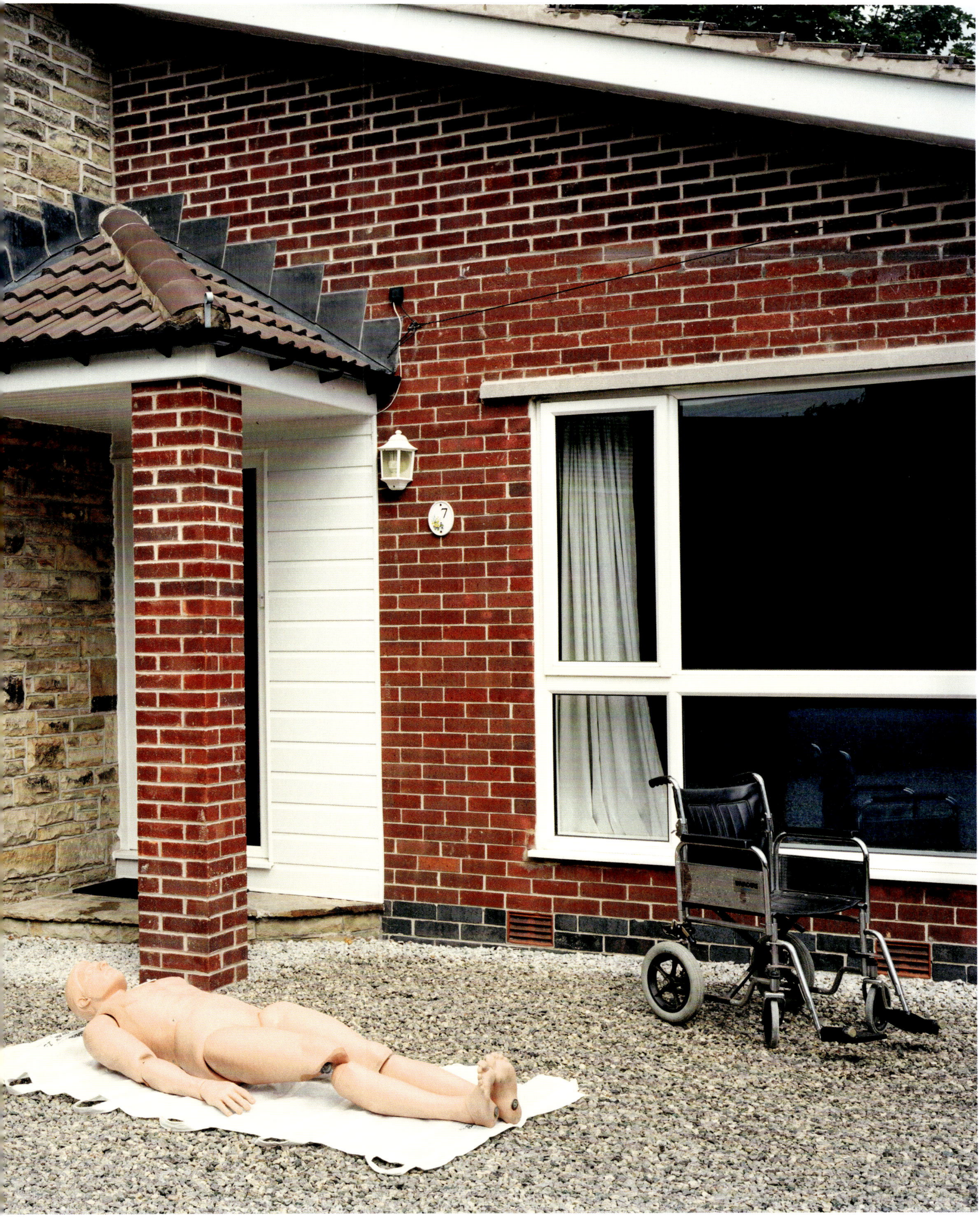
7

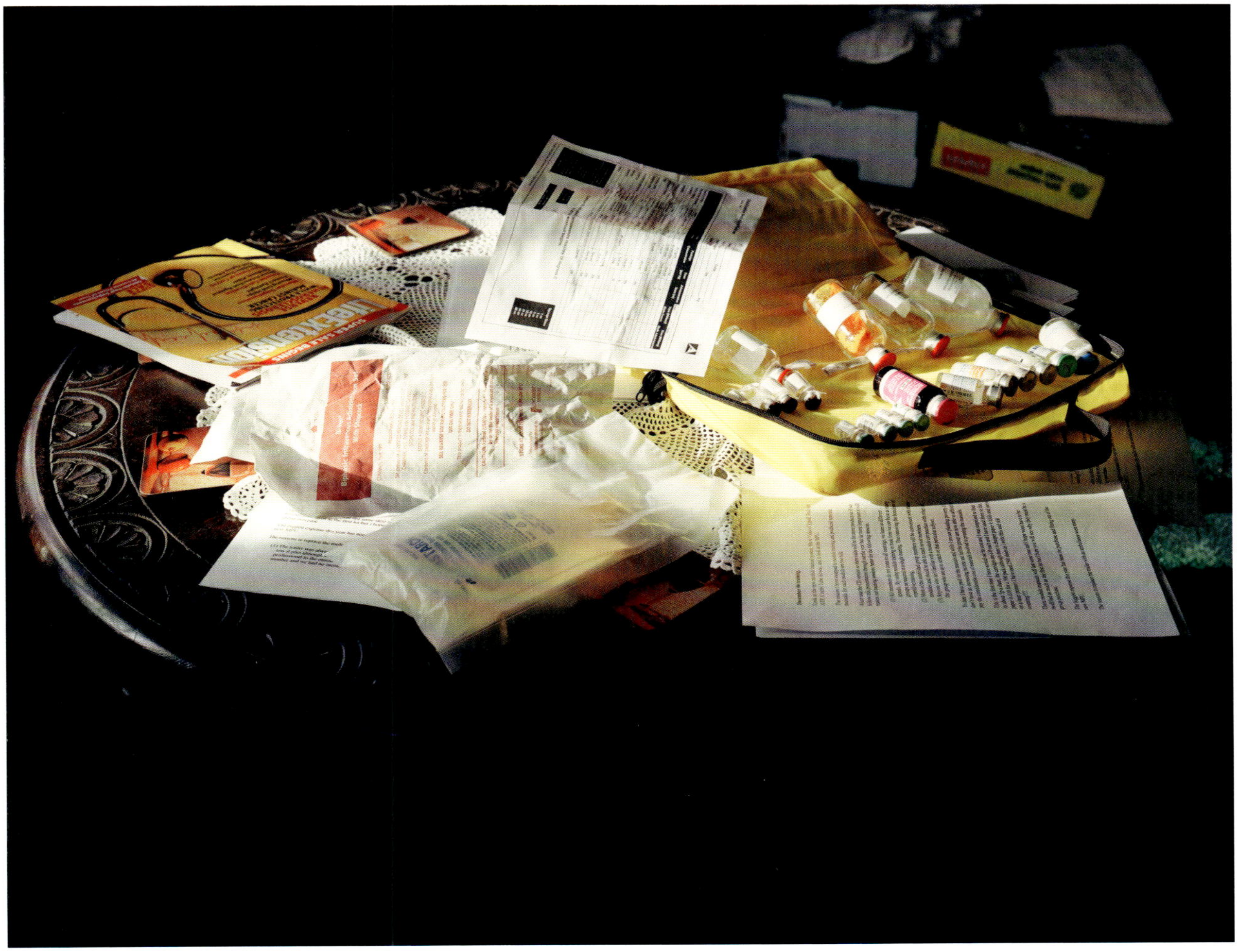

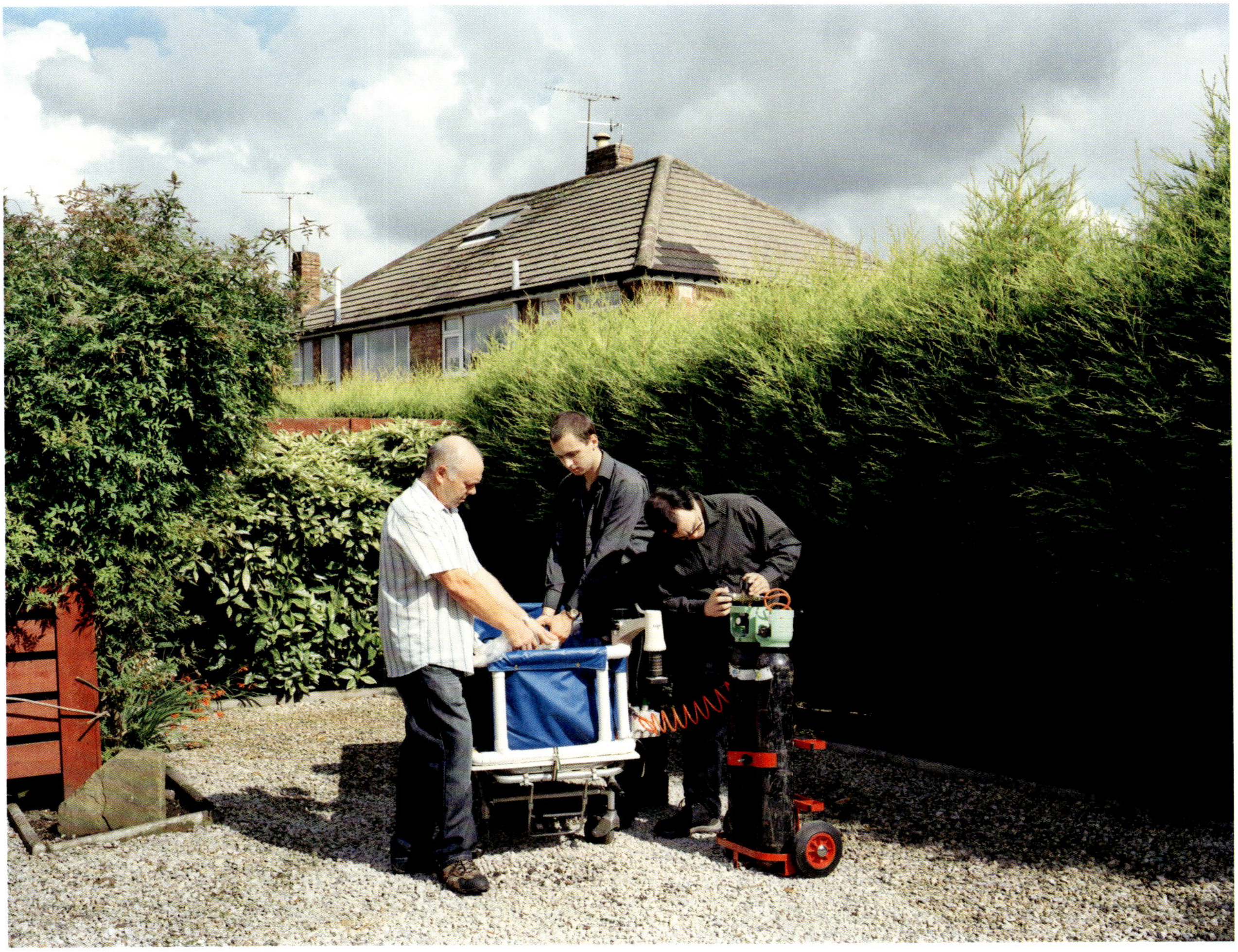

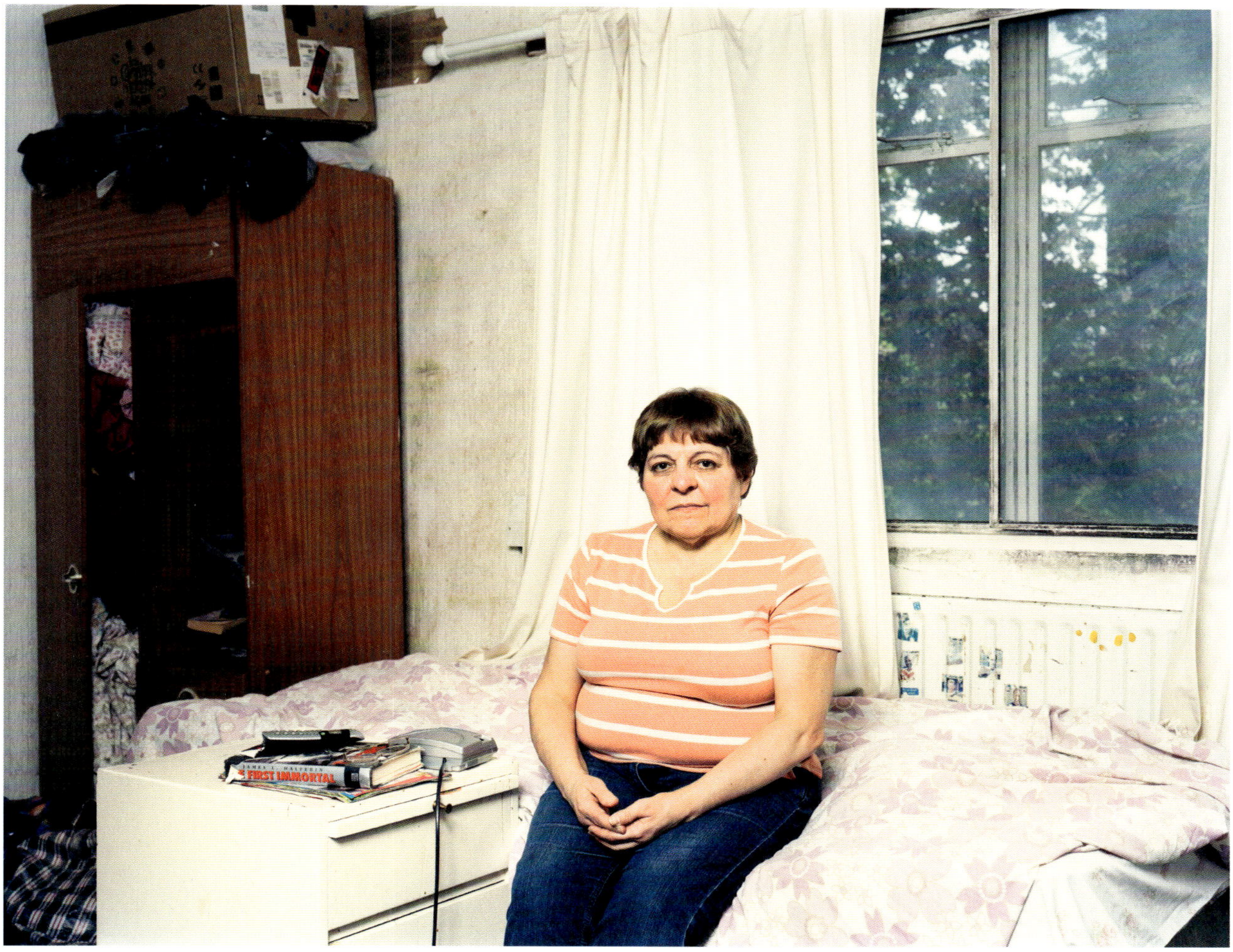

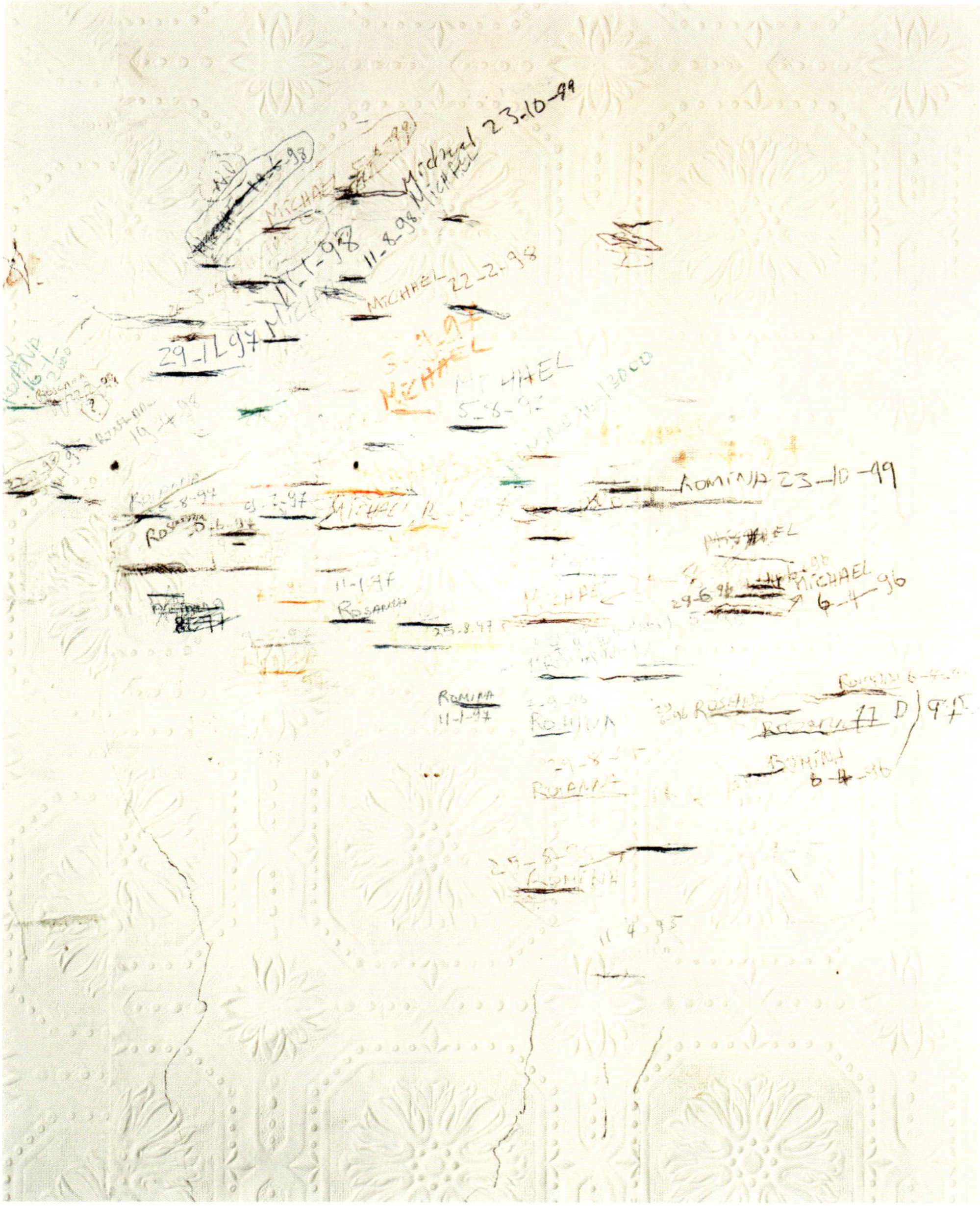
Michael 23-10-99
MICHAEL 22-2-98
11-8-98
29-1-97
ROMINA 23-10-99
MICHAEL 6-4-96

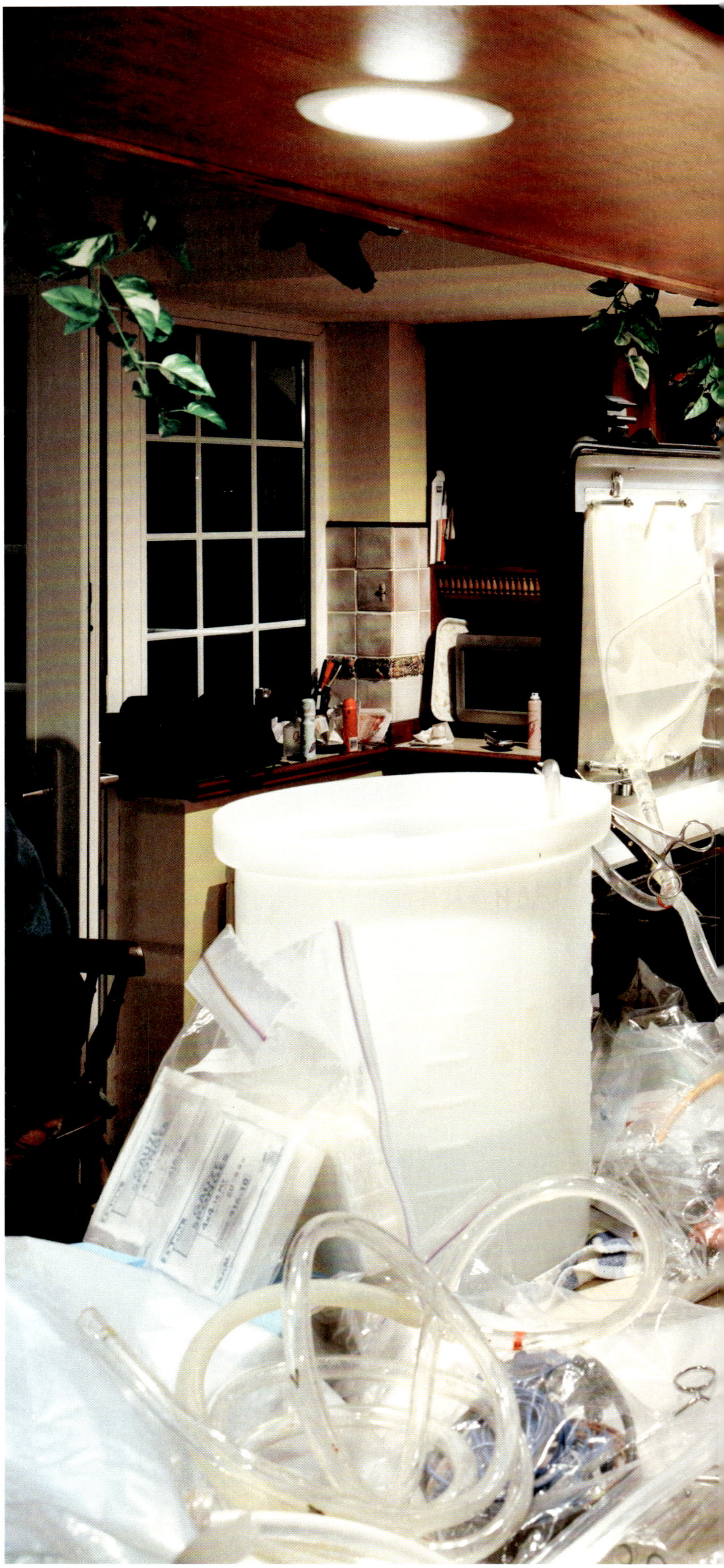

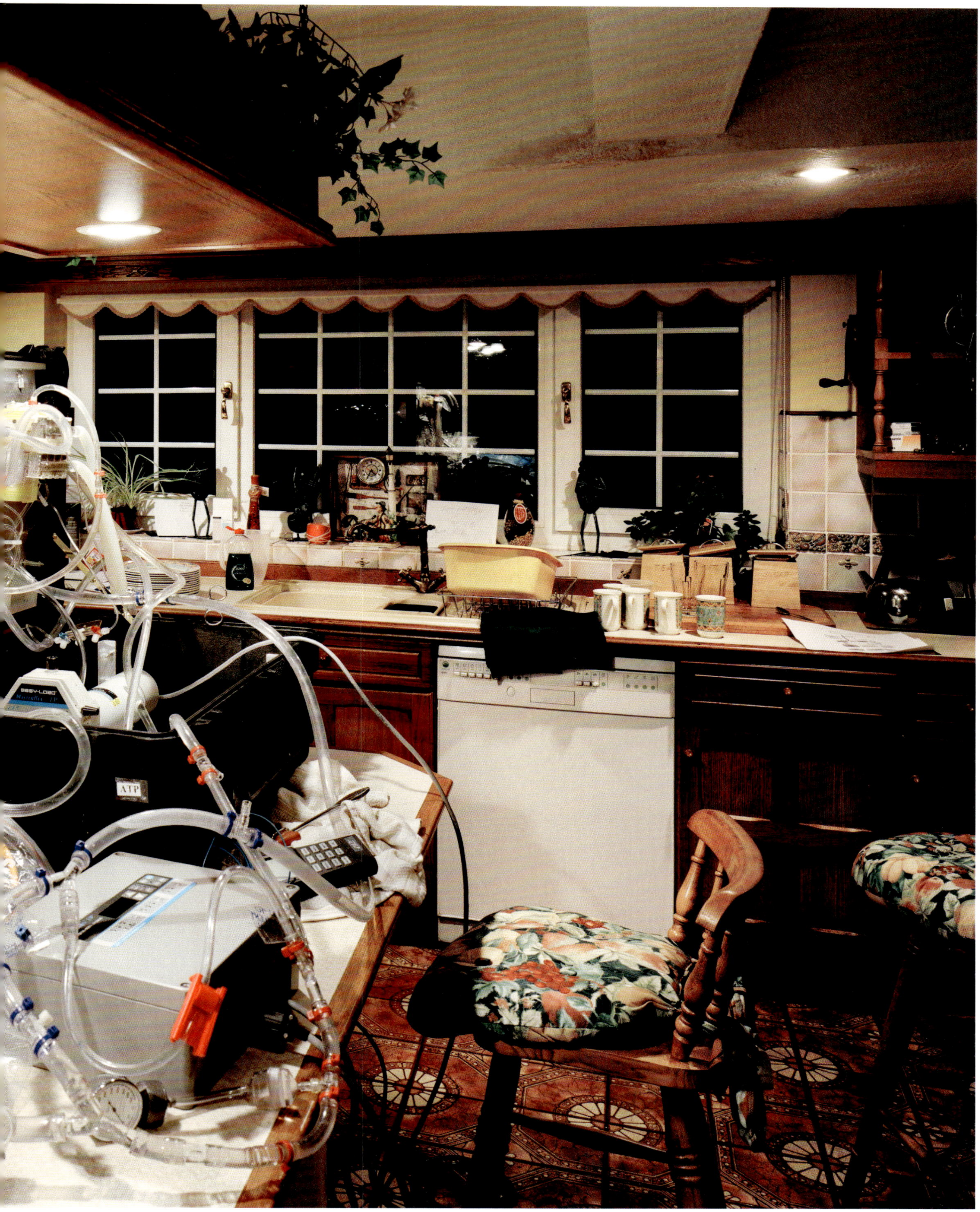
ATP

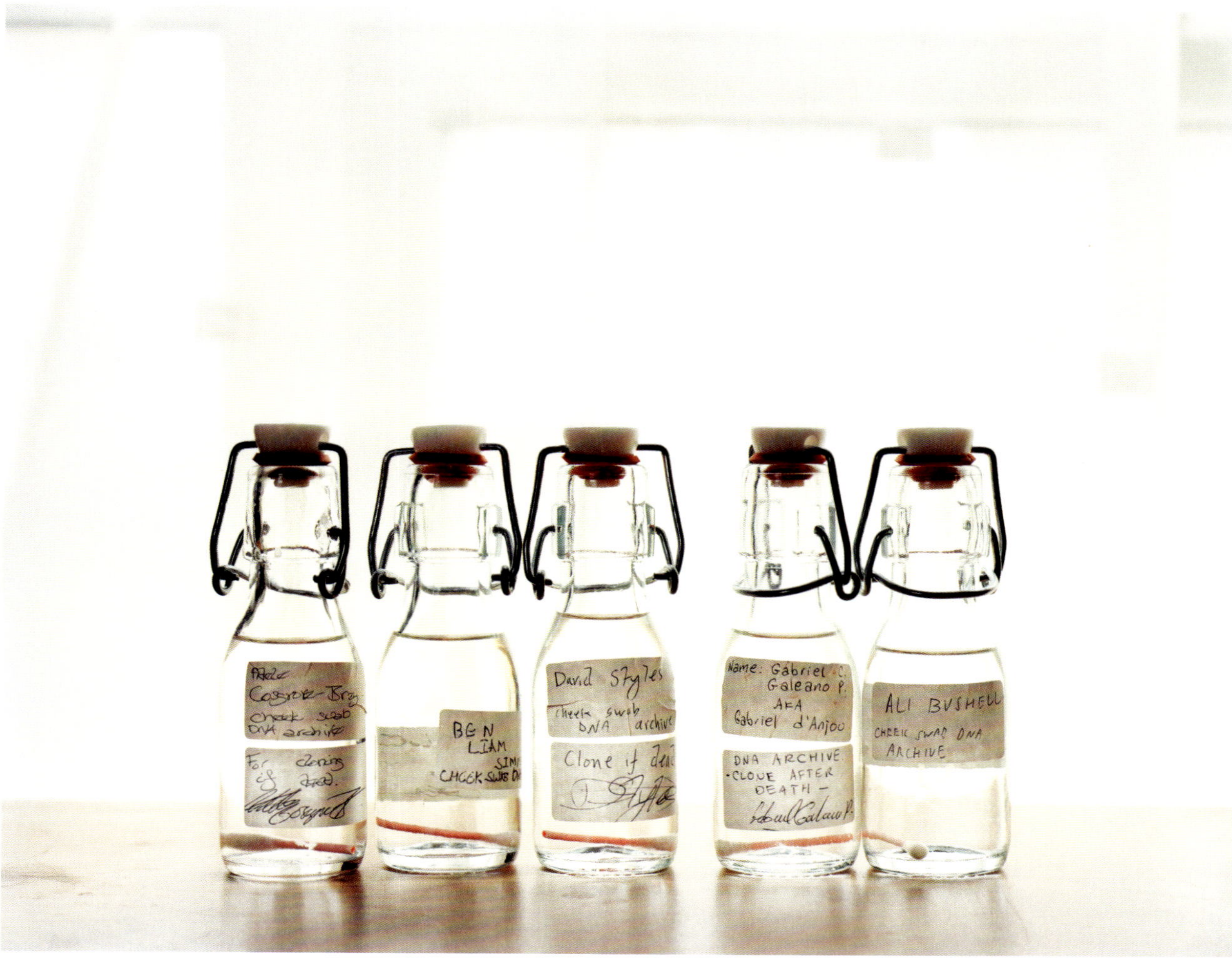
check swab DNA archive
For cloning if dead.
BEN LIAM
David Styles
cheek swab DNA archive
Name: Gabriel C. Galeano P.
AKA Gabriel d'Anjou
DNA ARCHIVE -CLONE AFTER DEATH -
ALI BUSHELL
CHEEK SWAB DNA ARCHIVE

KRIORUS PART 1

MOSCOW / ST PETERSBURG

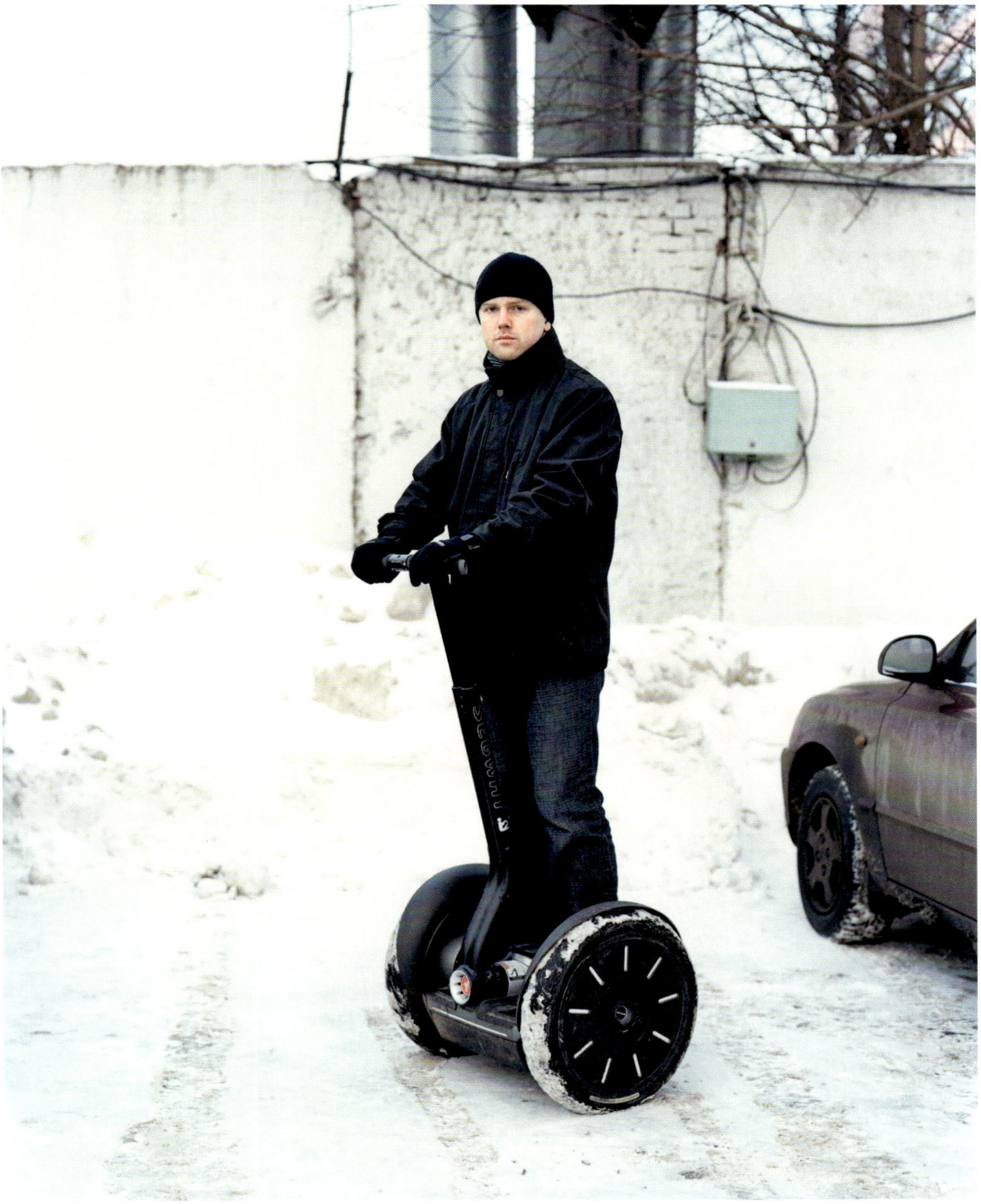

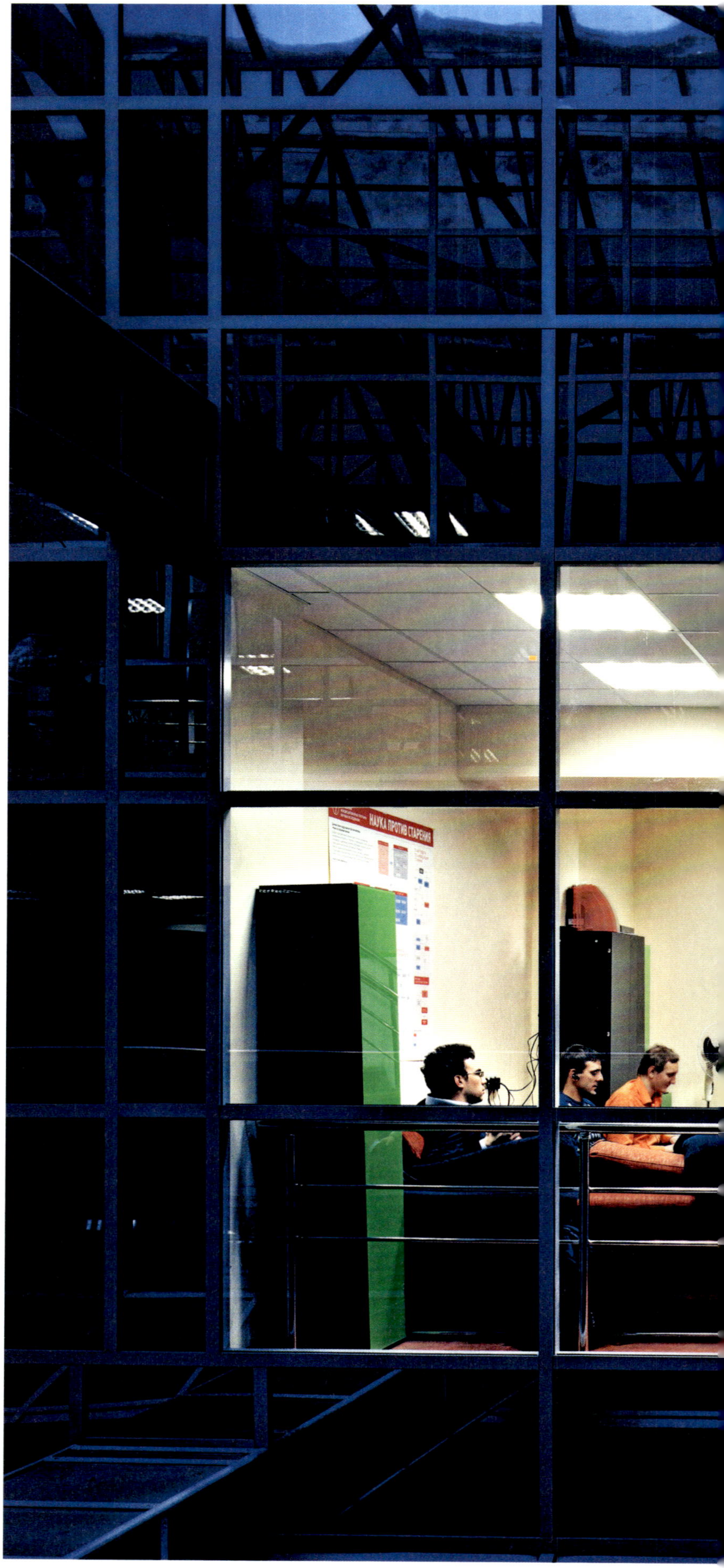
НАУКА ПРОТИВ СТАРЕНИЯ

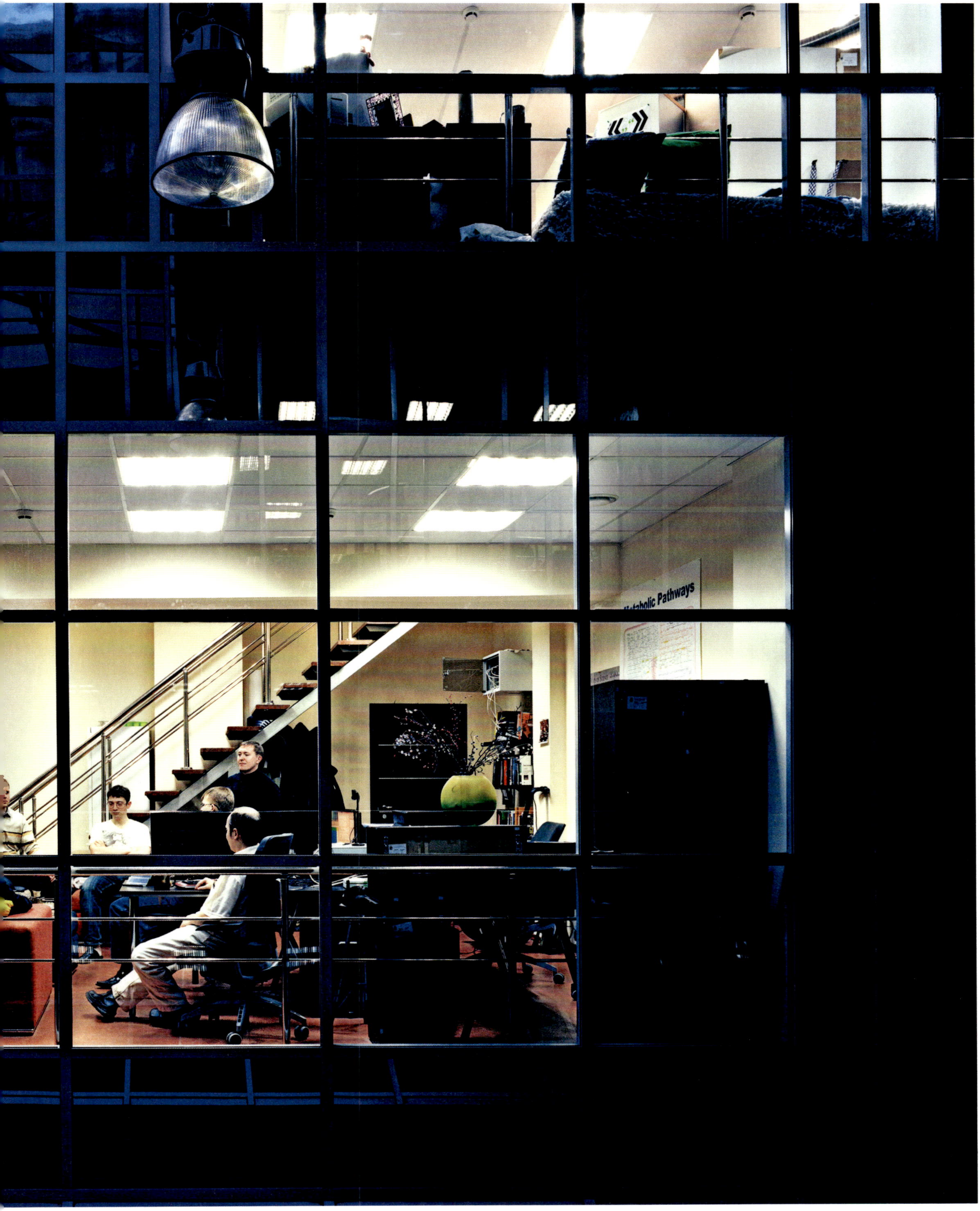
Pathways

By the summer of 2005, we understood what needed to be done in terms of establishing cryonics in Russia. We told the Cryonics Institute that we had formed a support group, and this group was made up of the people behind KrioRus.

I remember I wanted to write to a colleague, to tell him that we needed to have a set of instructions available for potential emergencies. When I got up from my bed in the middle of the night to write the email, I saw a message from the Cryonics Institute. They had been contacted by Daniil Fedorenko, who wanted to preserve his grandmother, Lidia.

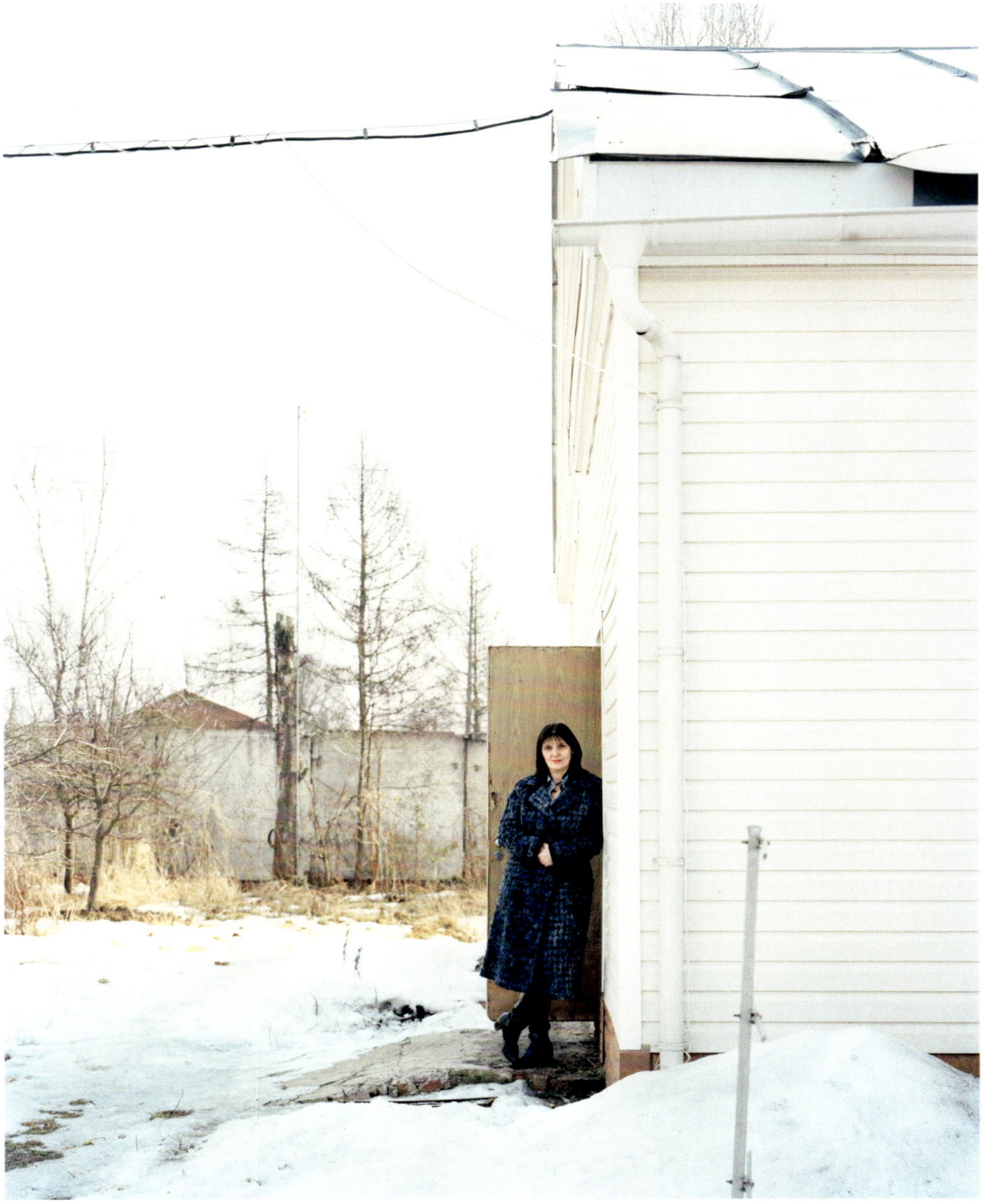

I immediately wrote to Daniil and arranged a meeting with him. I explained the situation – what was available, what we could do and what needed to be done, according to the understanding that we had about cryonics at that time. He said he was willing to do anything that it took to cryopreserve Lidia.

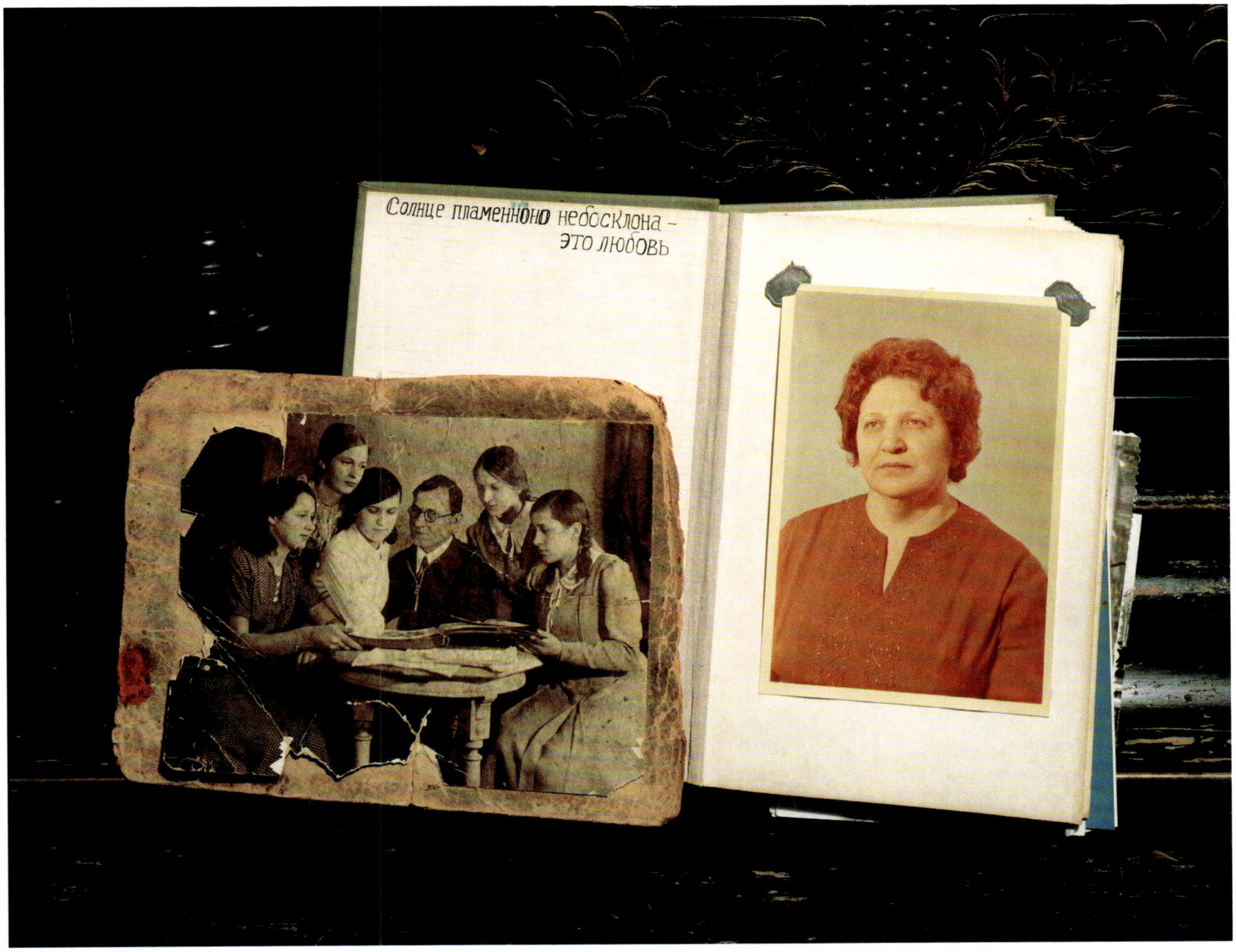
Солнце пламенного небосклона –
это любовь

Later I met Daniil with his mother, Galina, and we went to the hospital to see Lidia, who was still alive, but unconscious in the hospital bed. We started preparing Lidia for cryo-preservation.

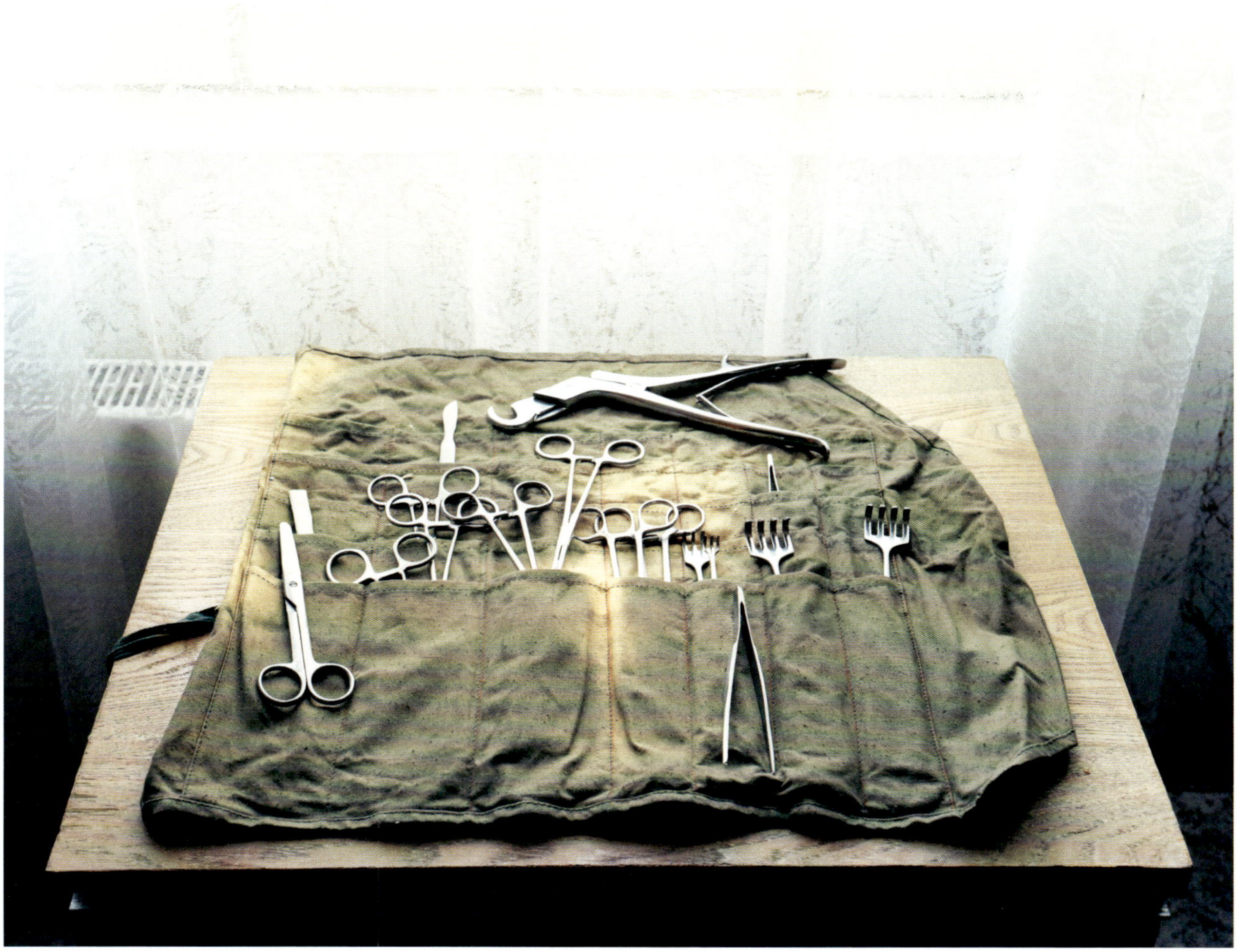

We didn't have any training, so we were very limited in what we were able to do. We knew the end result was that a person would be placed in liquid nitrogen or at a temperature low enough for the degradation process to be stopped. We had to act very fast and there was simply no time, so she was frozen using dry ice and was prepared for shipment to the United States.

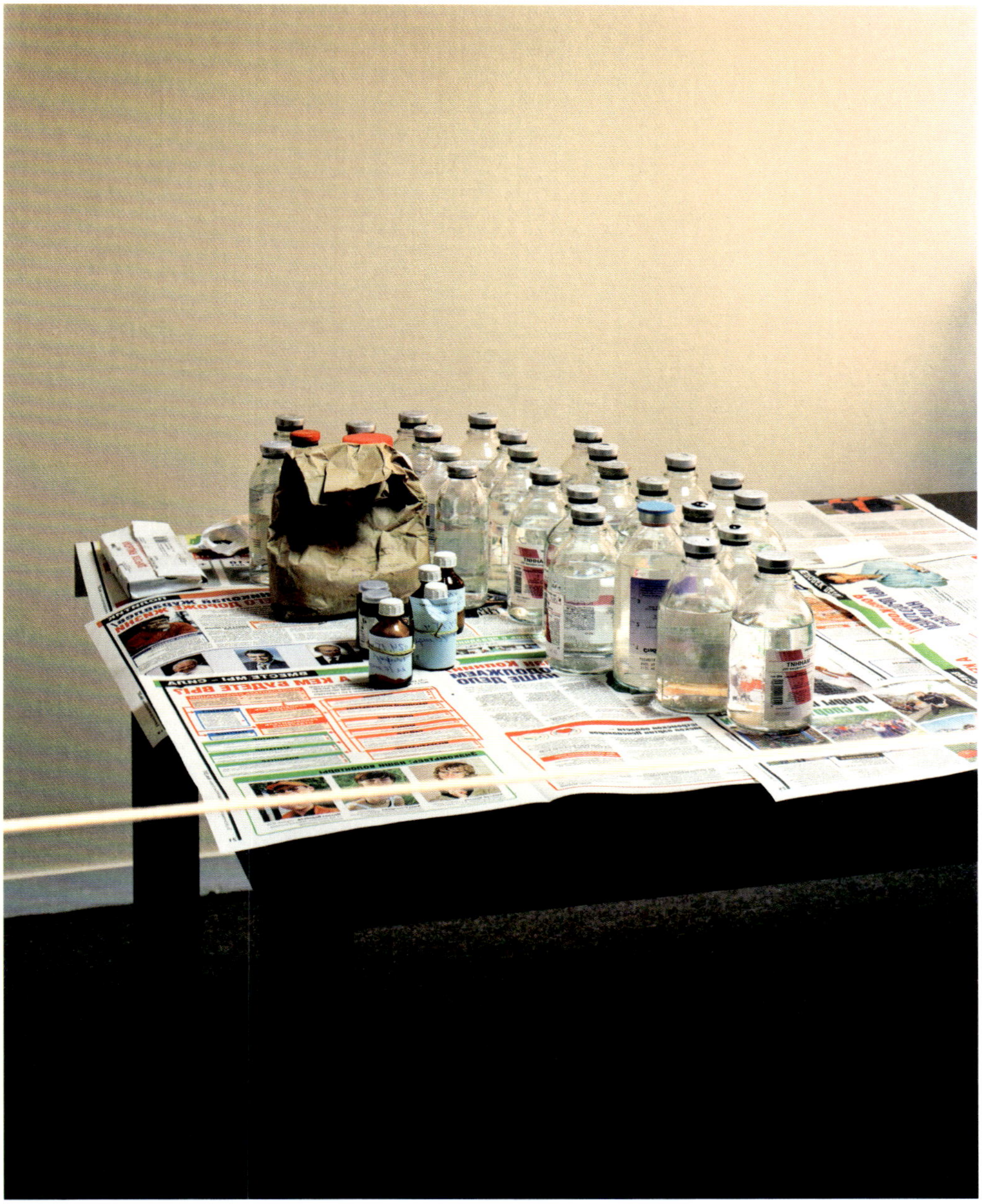

It was later when we realised that it would be prohibitively expensive to store a full body at the Cryonics Institute. So we stored Lidia here for a while. Her head is now in the KrioRus building and her body is buried in the cemetery.

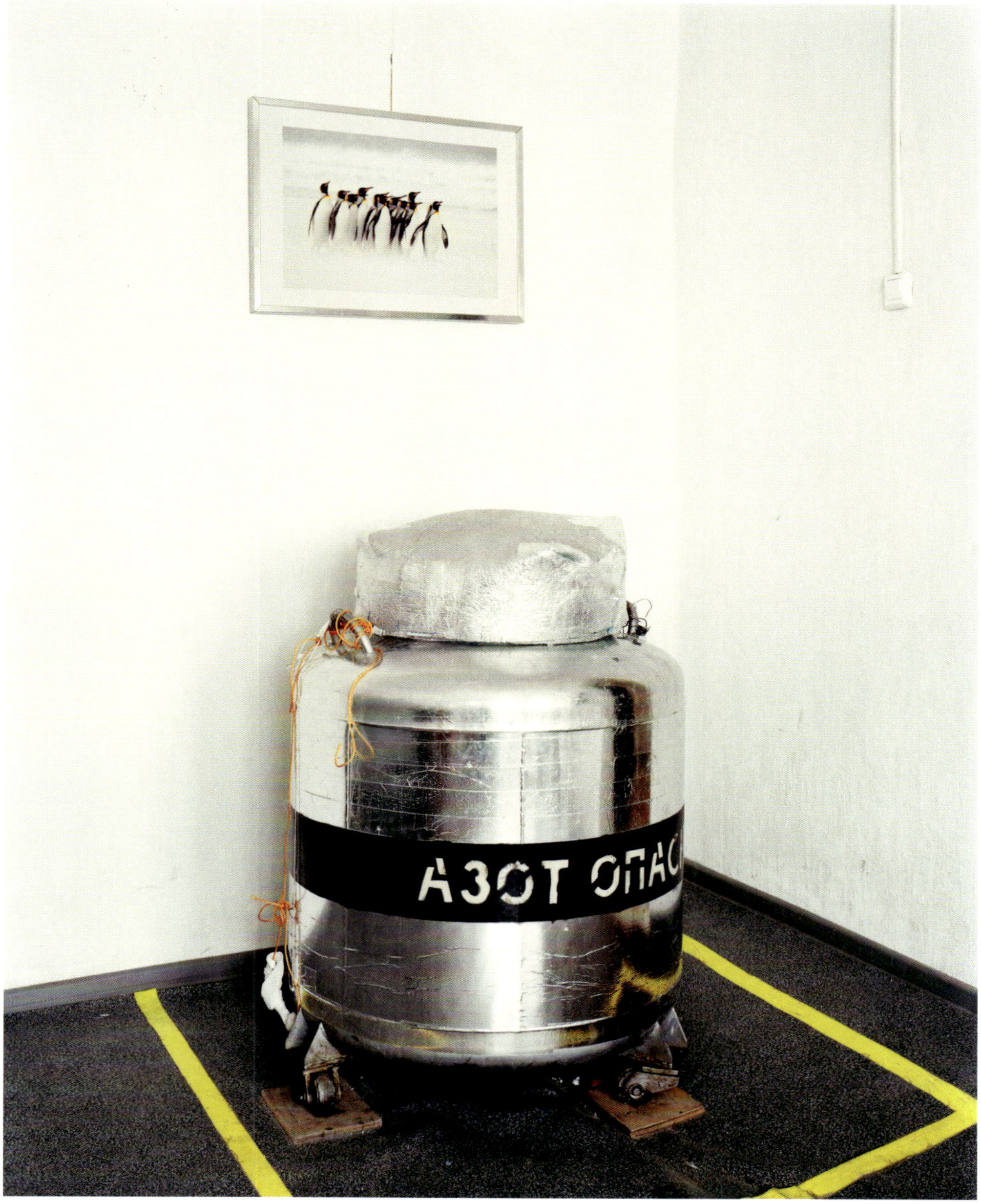
АЗОТ ОПАС

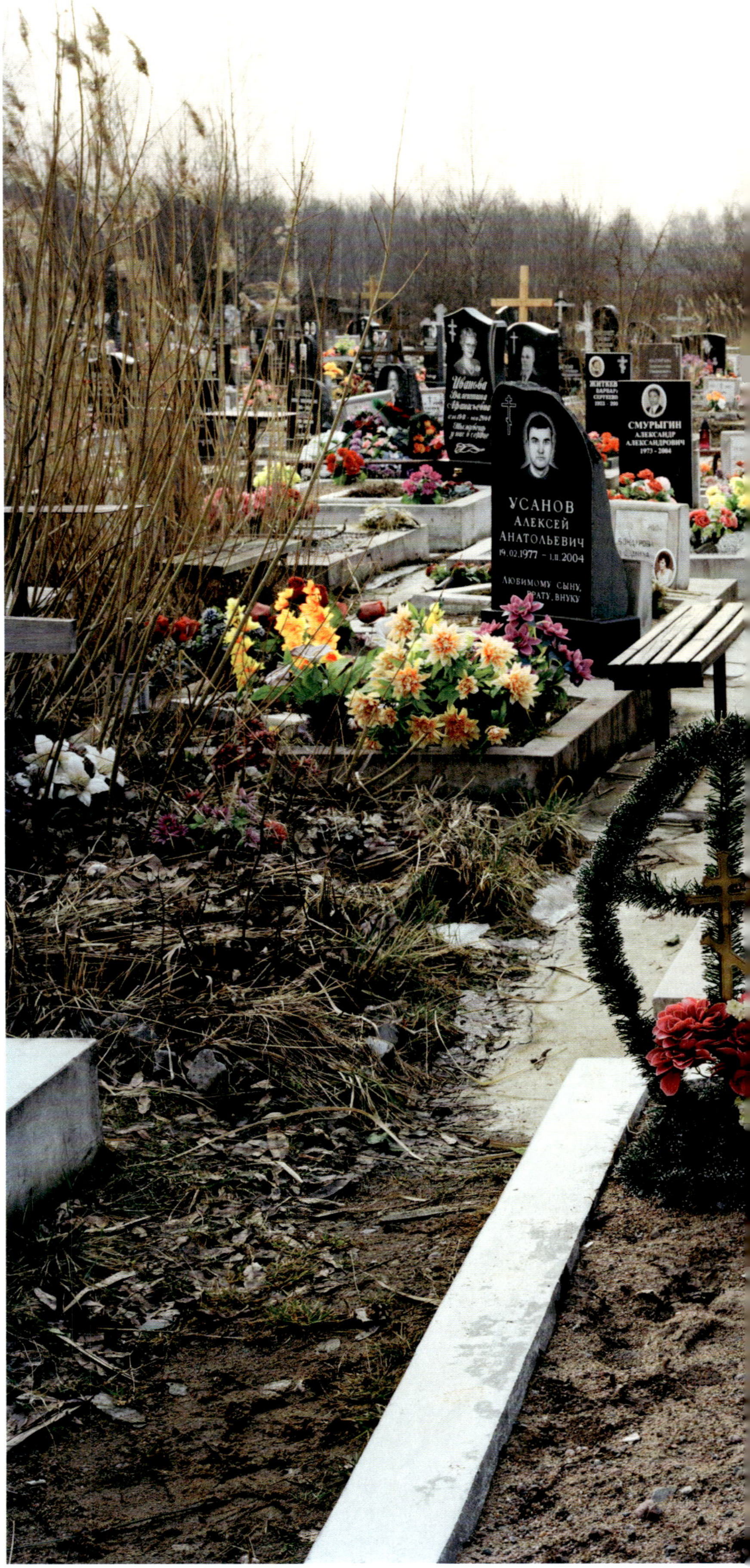
УСАНОВ
АЛЕКСЕЙ
АНАТОЛЬЕВИЧ
ЛЮБИМОМУ СЫНУ,
БРАТУ, ВНУКУ
СМУРЫГИН
АЛЕКСАНДР
АЛЕКСАНДРОВИЧ
1973 - 2004

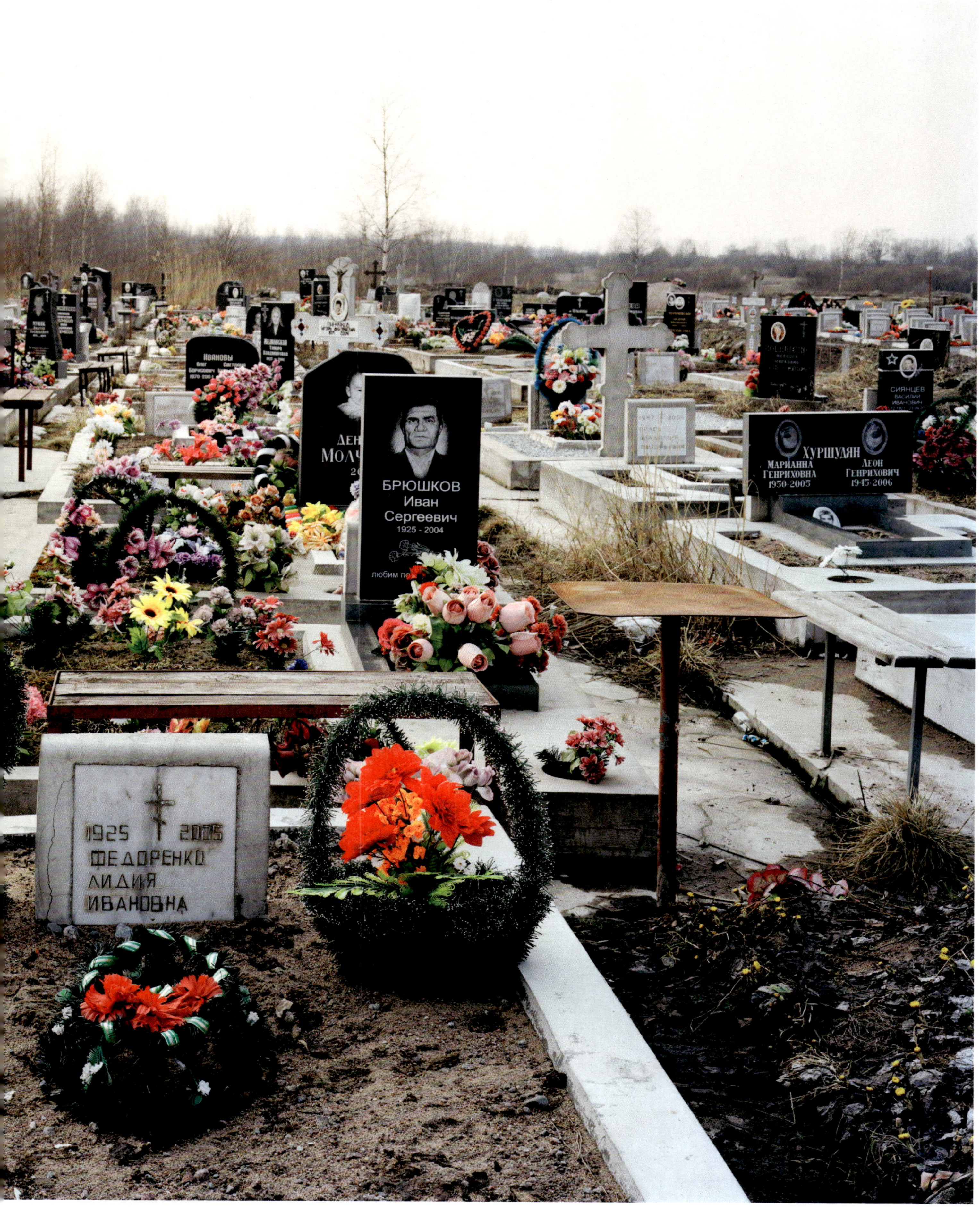

ИВАНОВЫ
БРЮШКОВ
Иван
Сергеевич
1925 - 2004
СИЯНЦЕВ
ХУРШУДЯН
МАРИАННА
ГЕНРИХОВНА
1950-2005
ЛЕОН
ГЕНРИХОВИЧ
1945-2006
1925 2005
ФЕДОРЕНКО
ЛИДИЯ
ИВАНОВНА

CRYONICS INSTITUTE

CLINTON TOWNSHIP, MICHIGAN

I have contracted with the Cryonics Institute to be cryonically preserved in the event of my death. It is my wish that the following be done immediately after clinical death.

Number One – Cool me by topical application of ice or ice packs with special attention to the head. If the ice is directly from the freezer, use a towel between the patient and the ice to avoid surface freezing.

Number Two – Please leave in place any access or drainage tools such as IV, NG, Foley and/or tracheotomy tool with oxygen line.

Number Three – Administer 30,000 units of heparin, intravenously if possible.

Number Four – If possible, after pronouncement of death, use prompt CPR-like compressions to maintain heart and lung function. Vigorous compressions from four to 15 minutes help to delay deterioration, improve heat transfer and help to circulate heparin throughout the body. Circulation of heparin is of greater urgency than the risk of broken ribs.

Number Five – Call Suspended Animation at Boynton Beach, Florida, and report the death. I also list two phone numbers to be called.

Number Six – Call the Cryonics Institute emergency telephone numbers until Cryonics Institute personnel are alerted. I list two phone numbers for them.

Number Seven – Release the patient to Suspended Animation personnel as soon as possible, without autopsy.

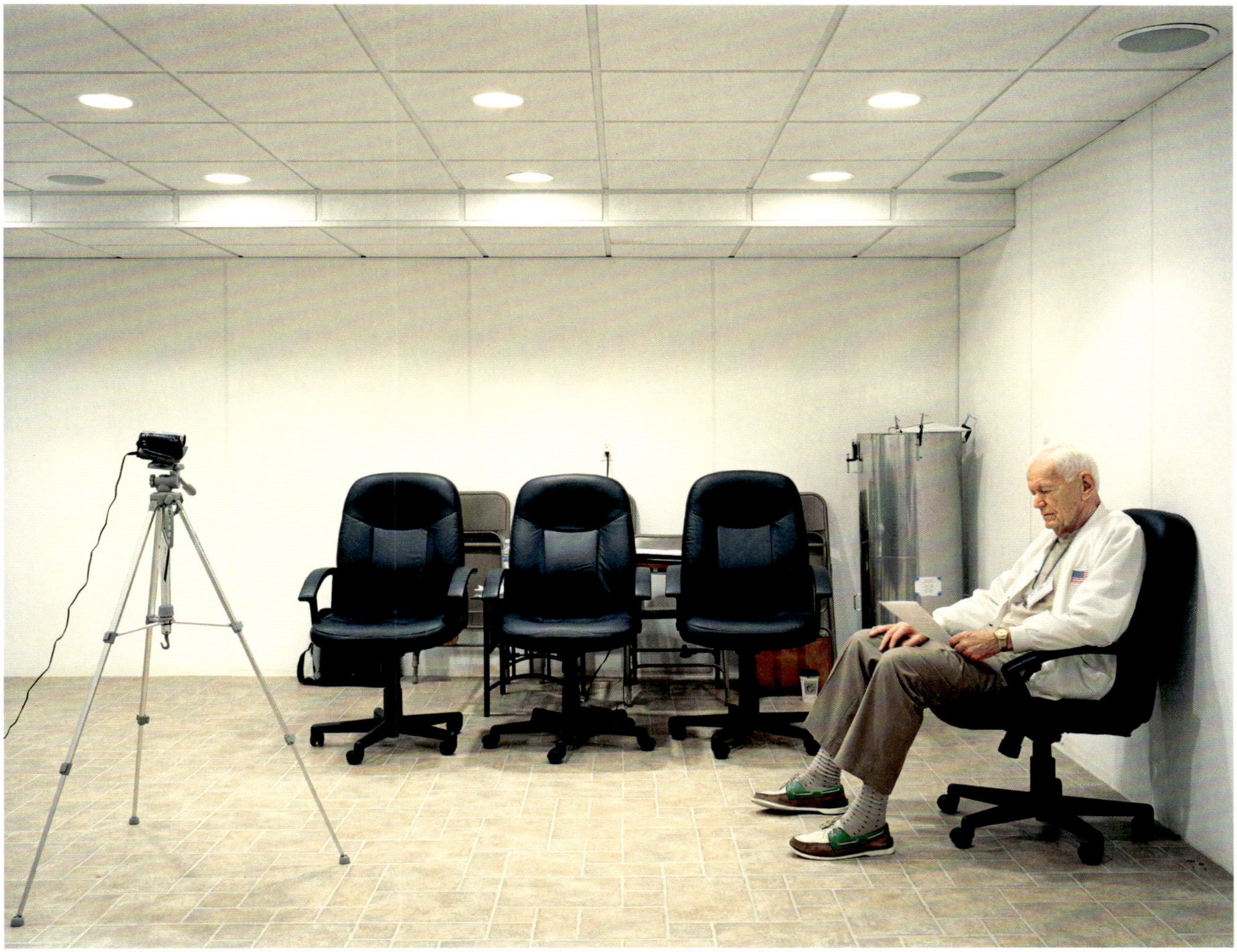

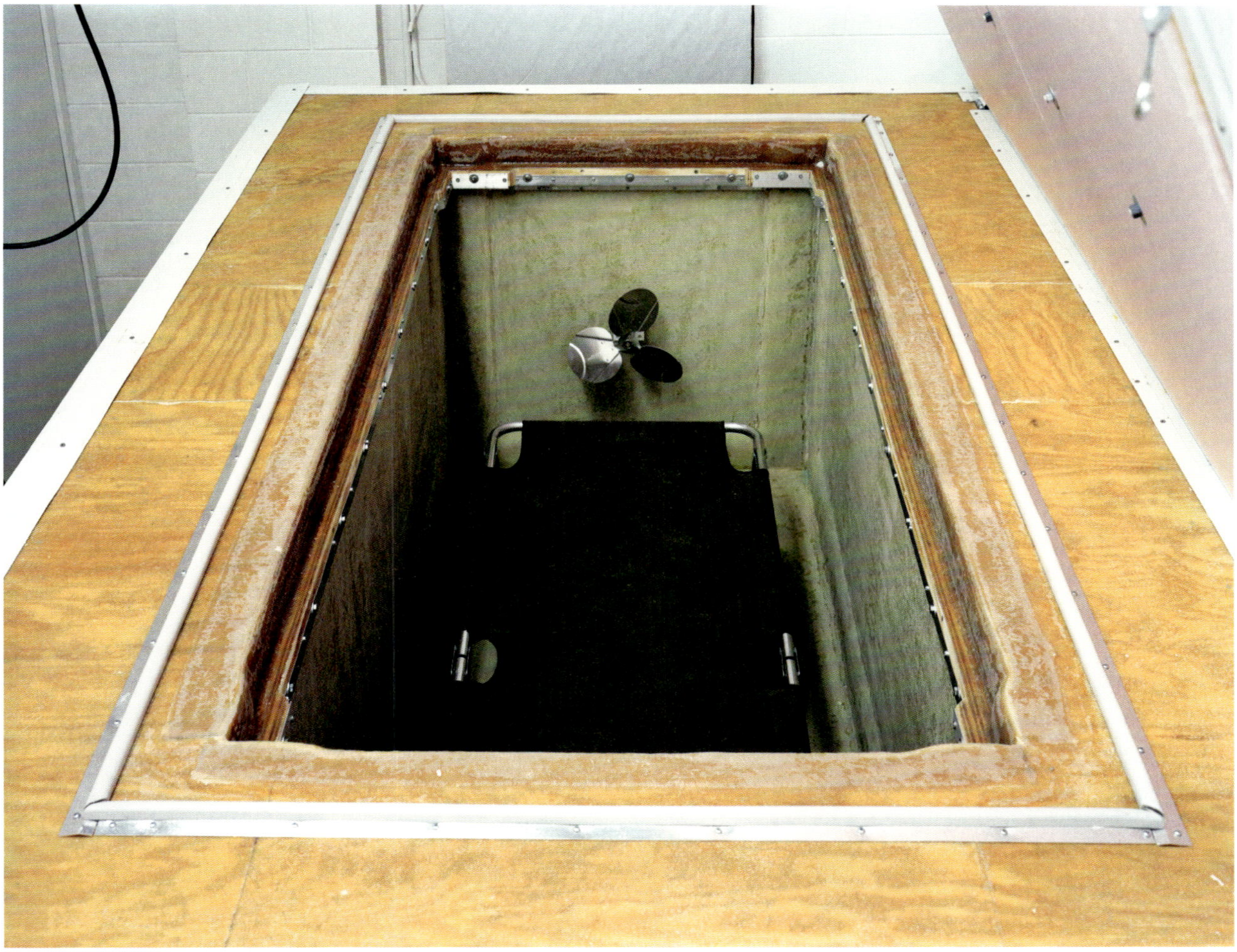

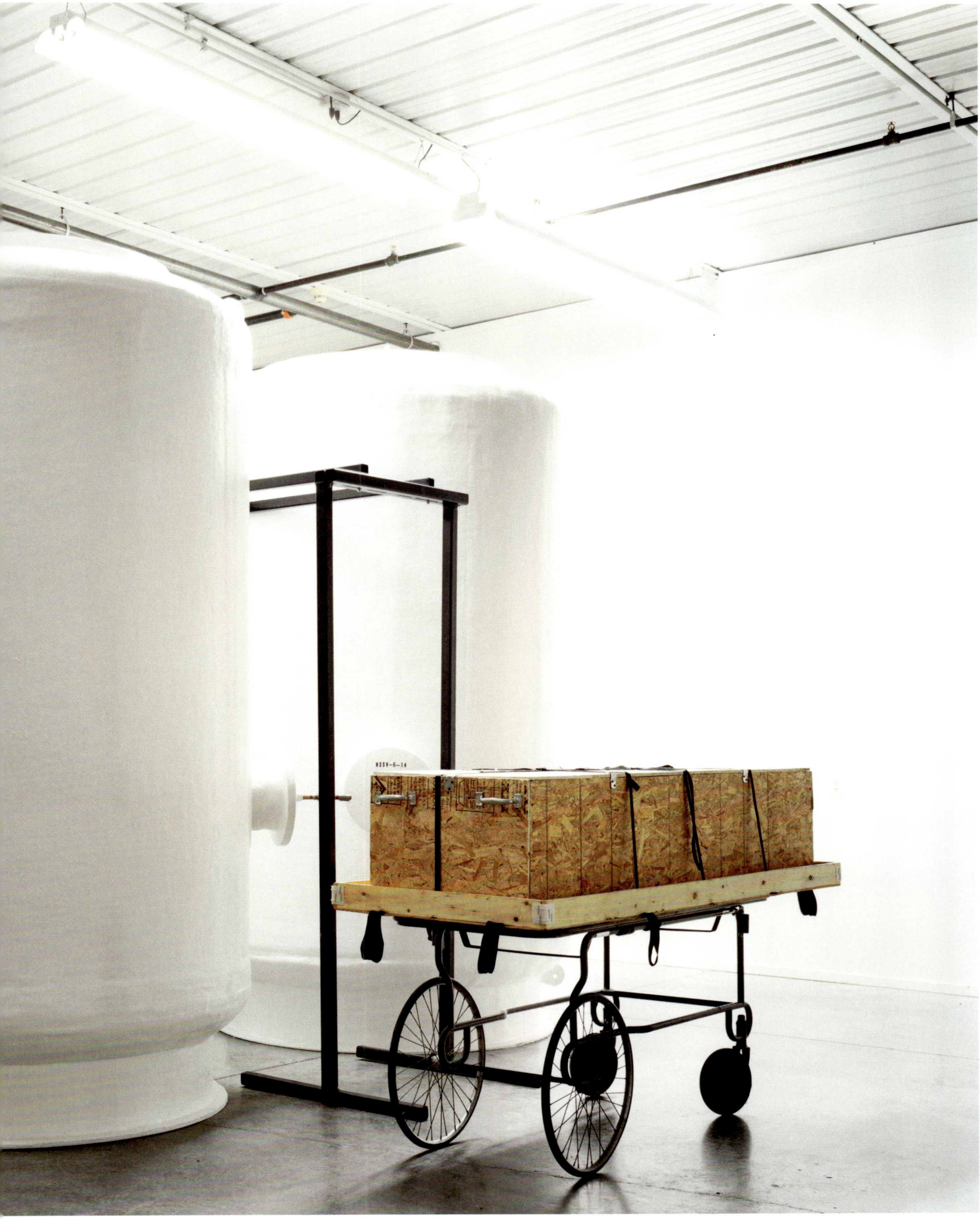
HSSV-6-14

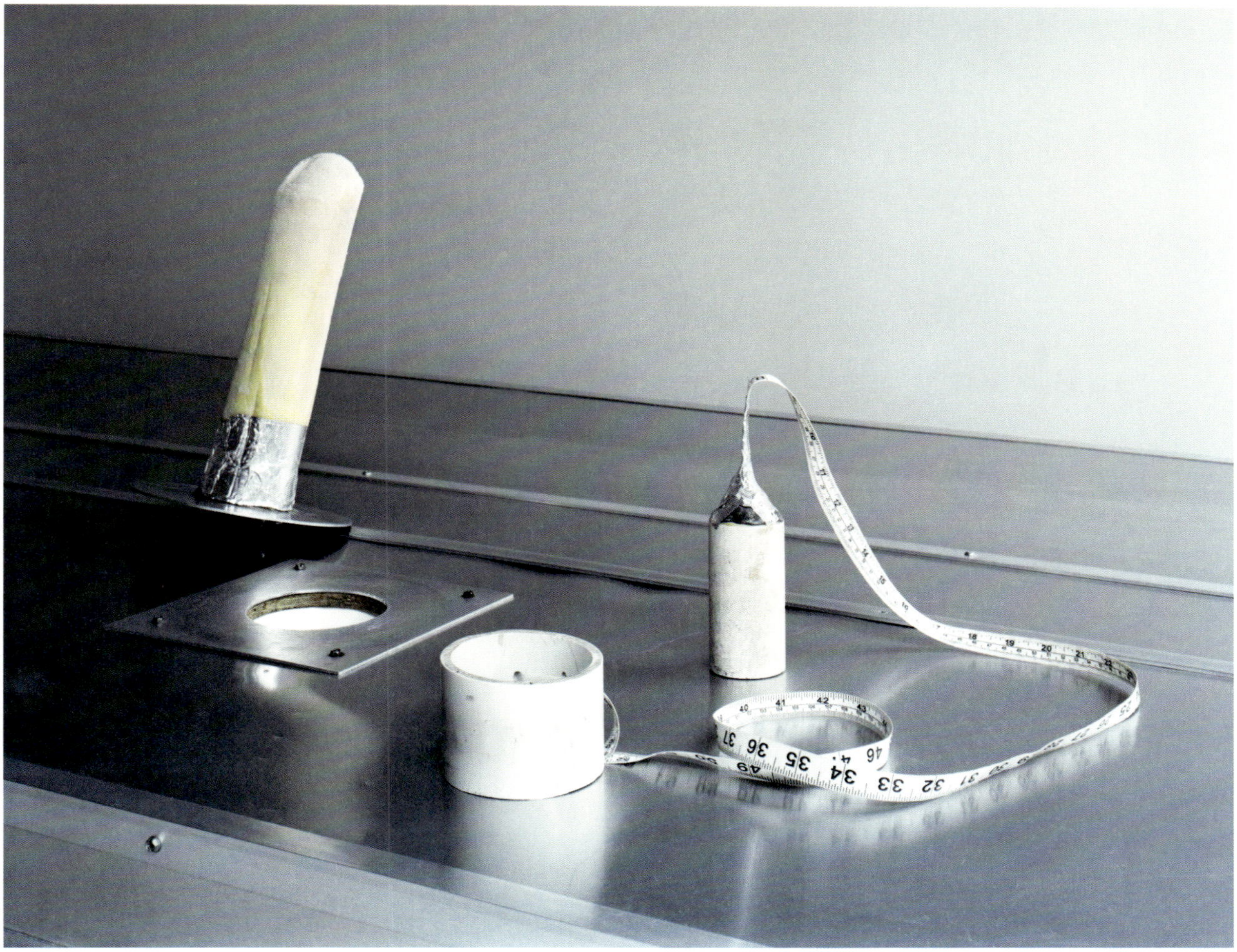

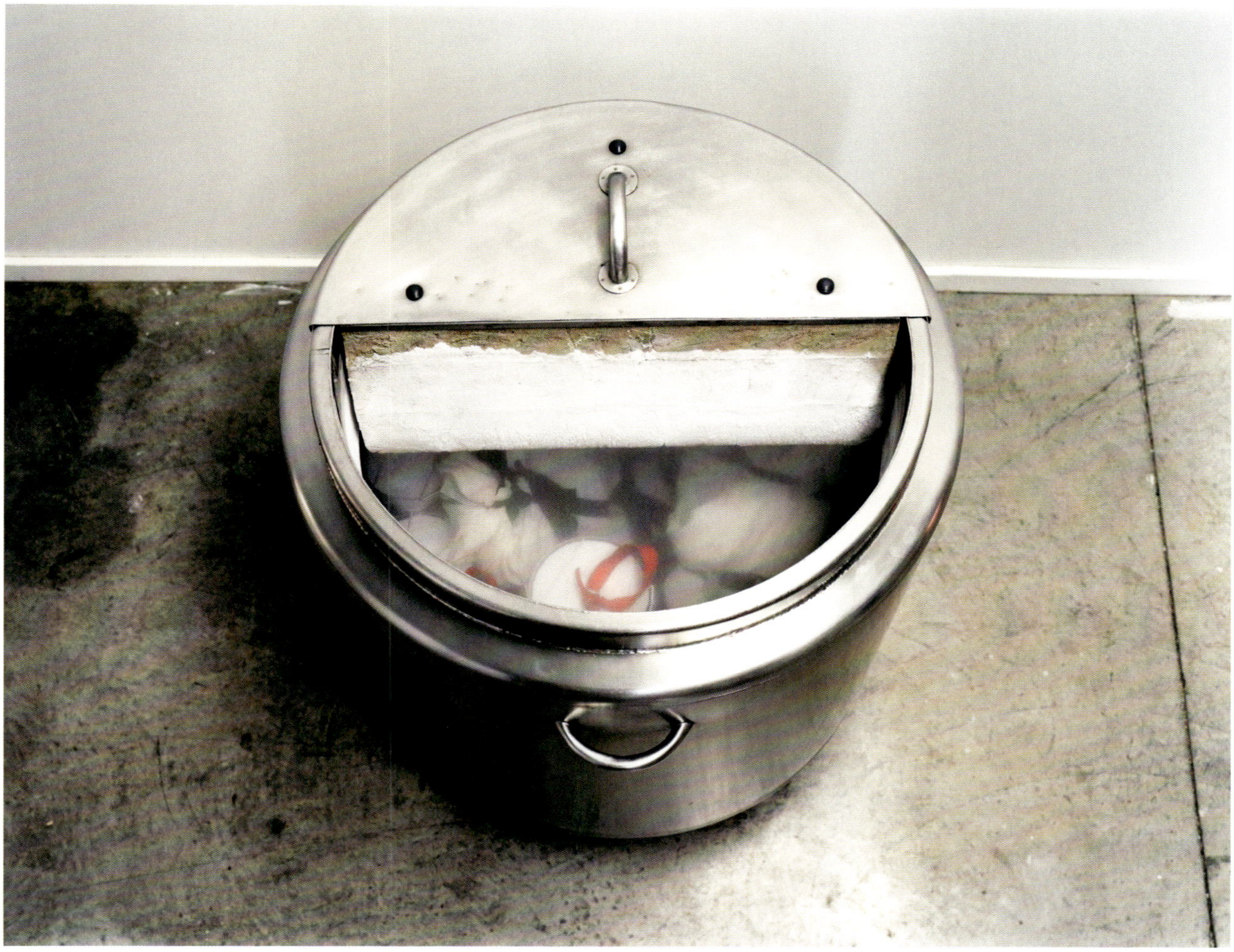

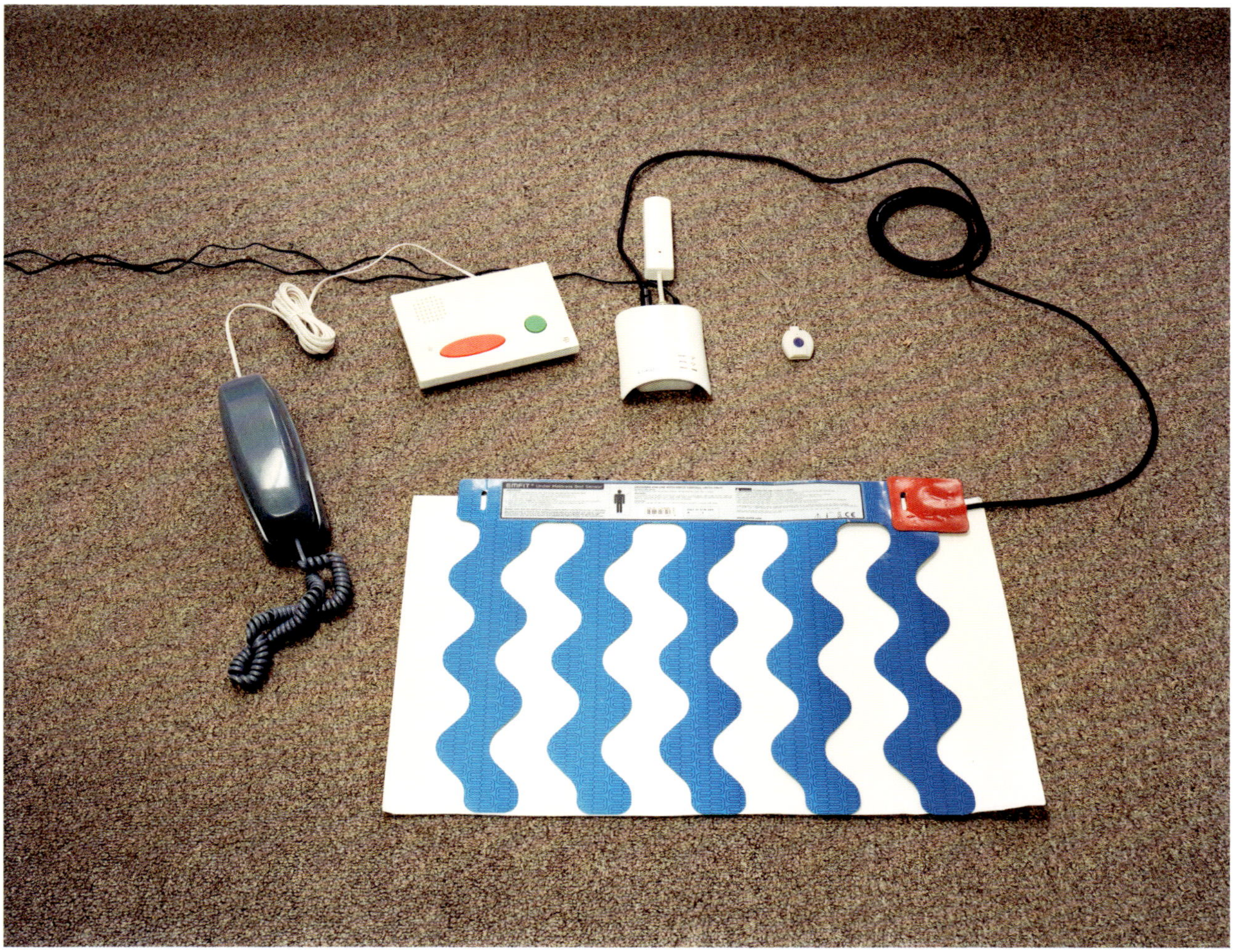

A PLACE OF LAST STAY

ROSEVILLE, MICHIGAN

My name is George Vogler and I've been with the cryonics movement since the late 1960s. Recently I purchased a home close to the Cryonics Institute so that when my end comes in this life, or in this period of life, I'll be nearby.

It's my intention to set the house up as a place of last stay for people coming into the state from other places. They'll have surroundings that are comfortable and will not have to worry about the usual problems arising from trying to live in a hotel or a hospital or something of this sort.

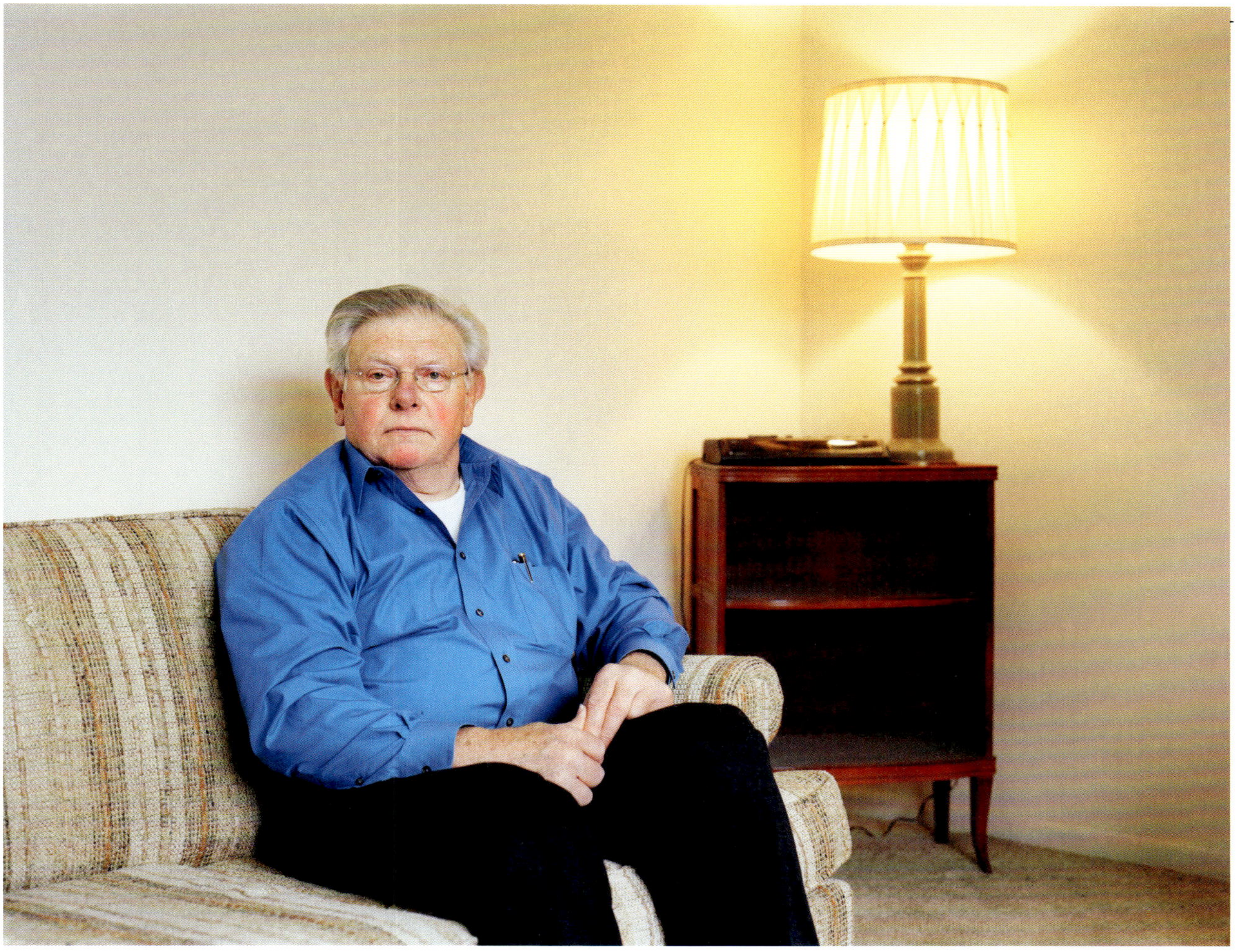

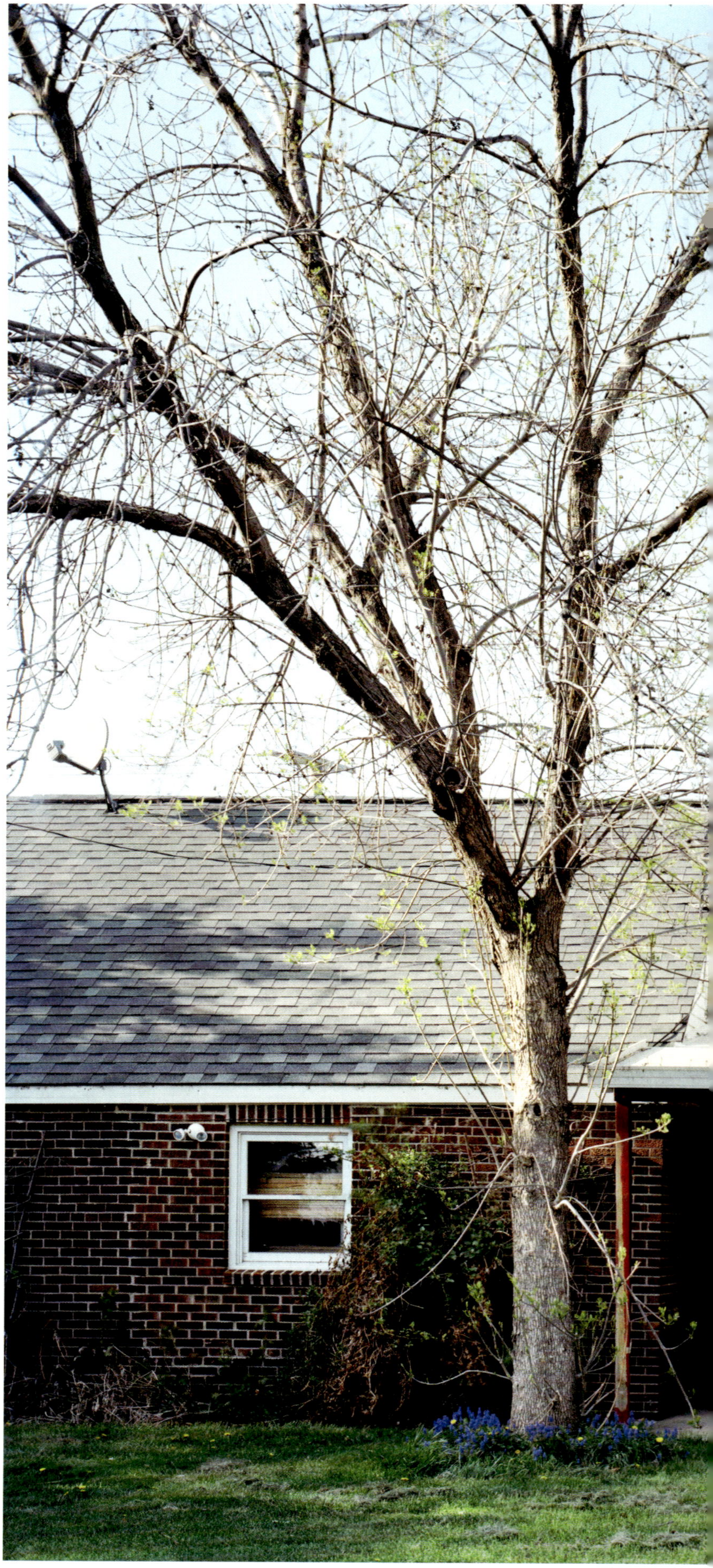

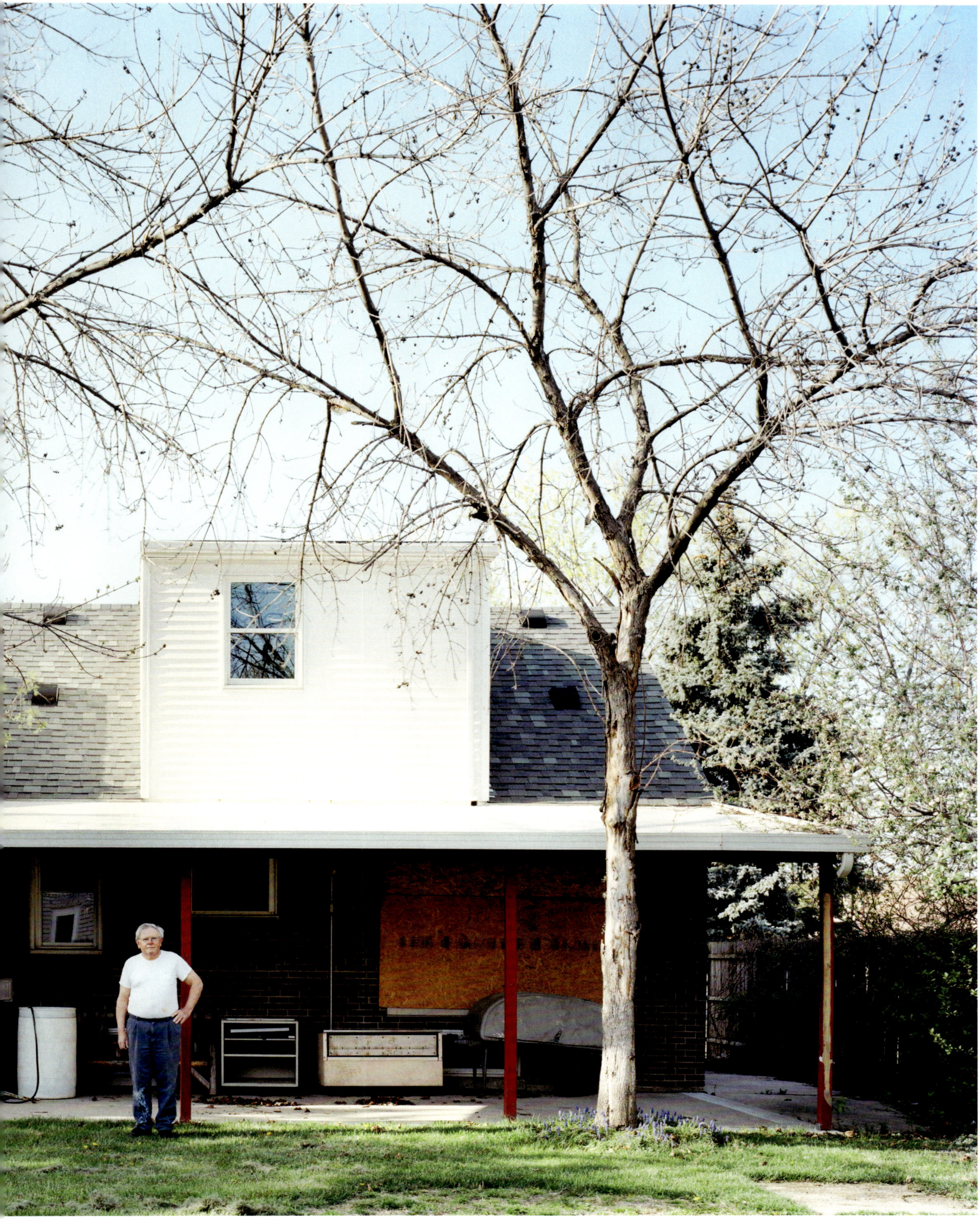

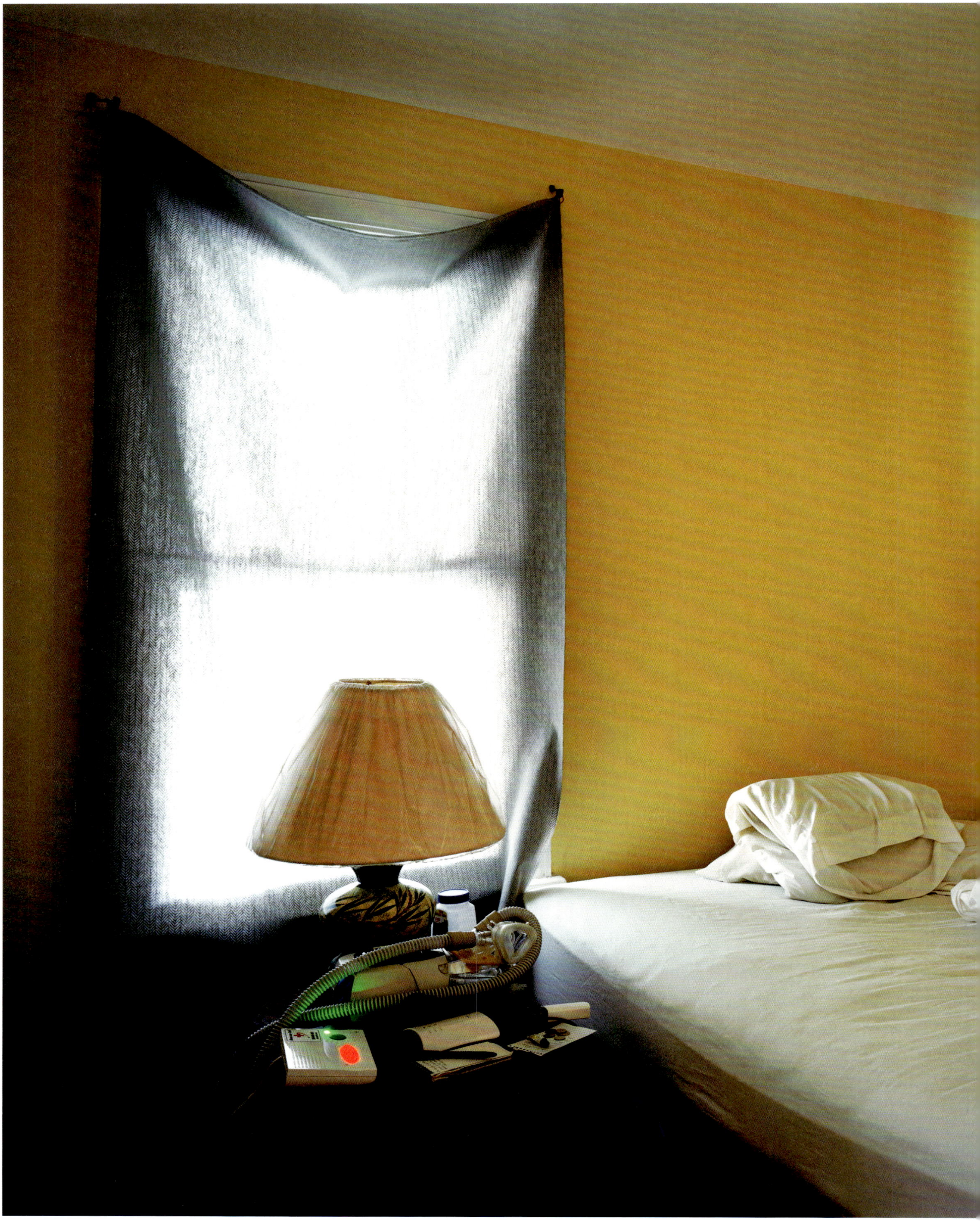

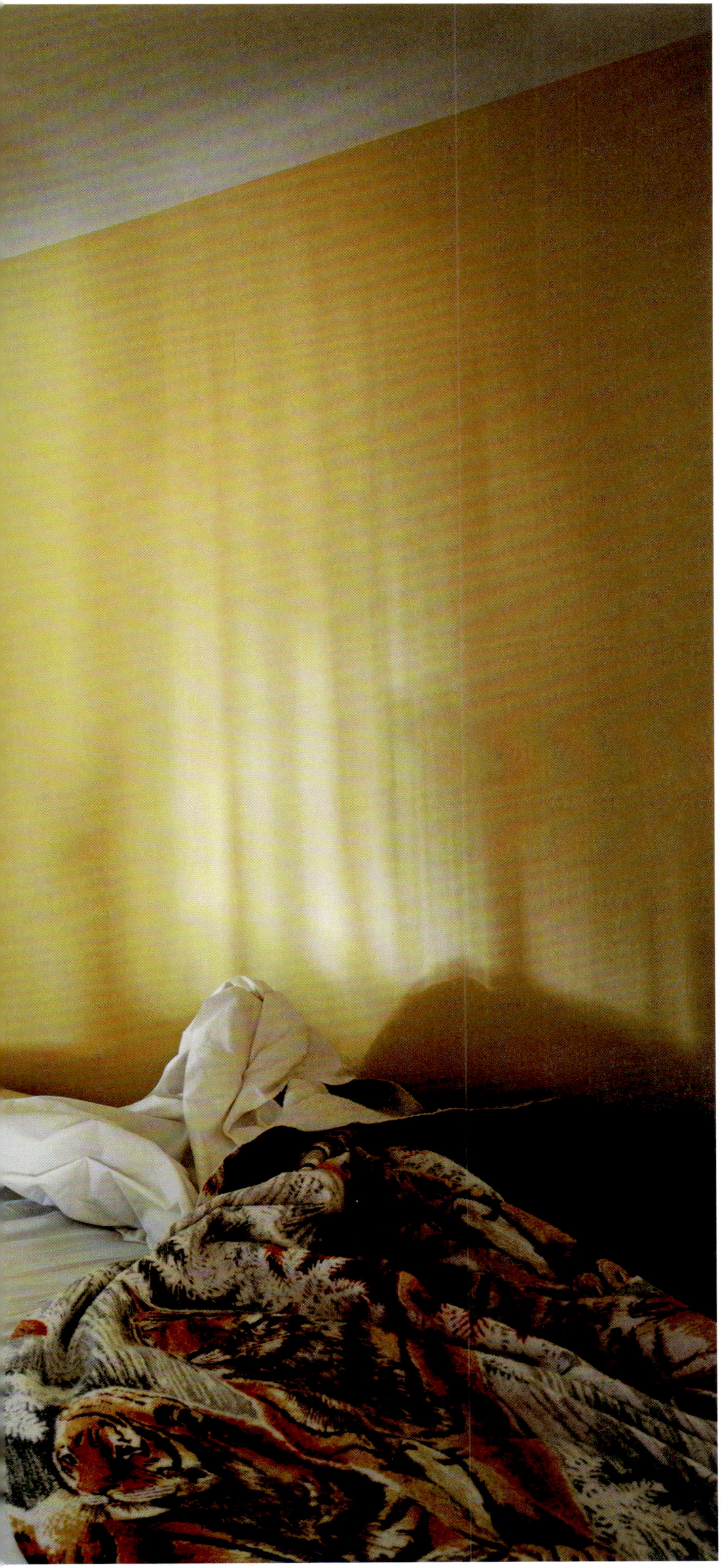

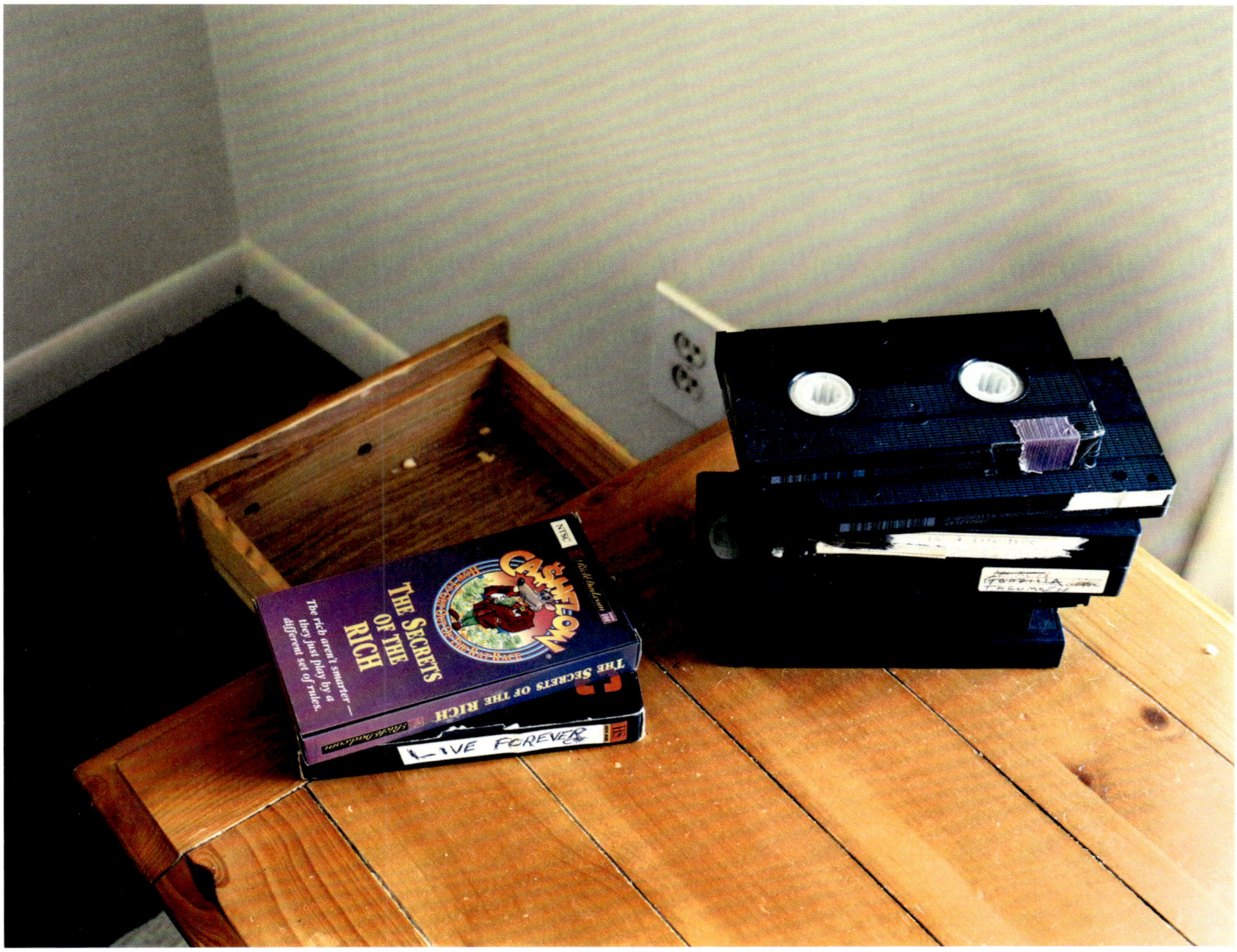
NTSC
CASHFLOW
HOW TO GET OUT OF THE RAT RACE
THE SECRETS OF THE RICH
The rich aren't smarter — they just play by a different set of rules.
THE SECRETS OF THE RICH
LIVE FOREVER

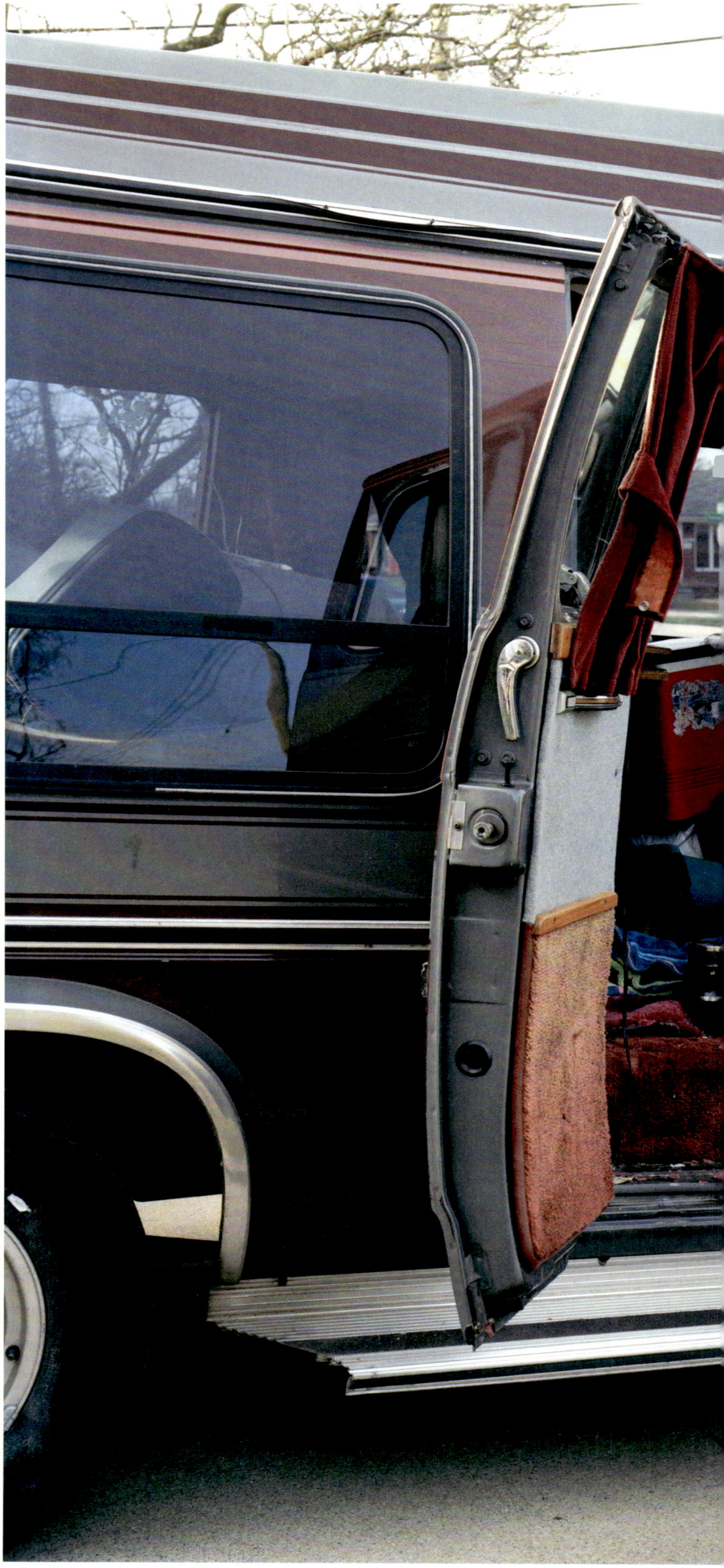

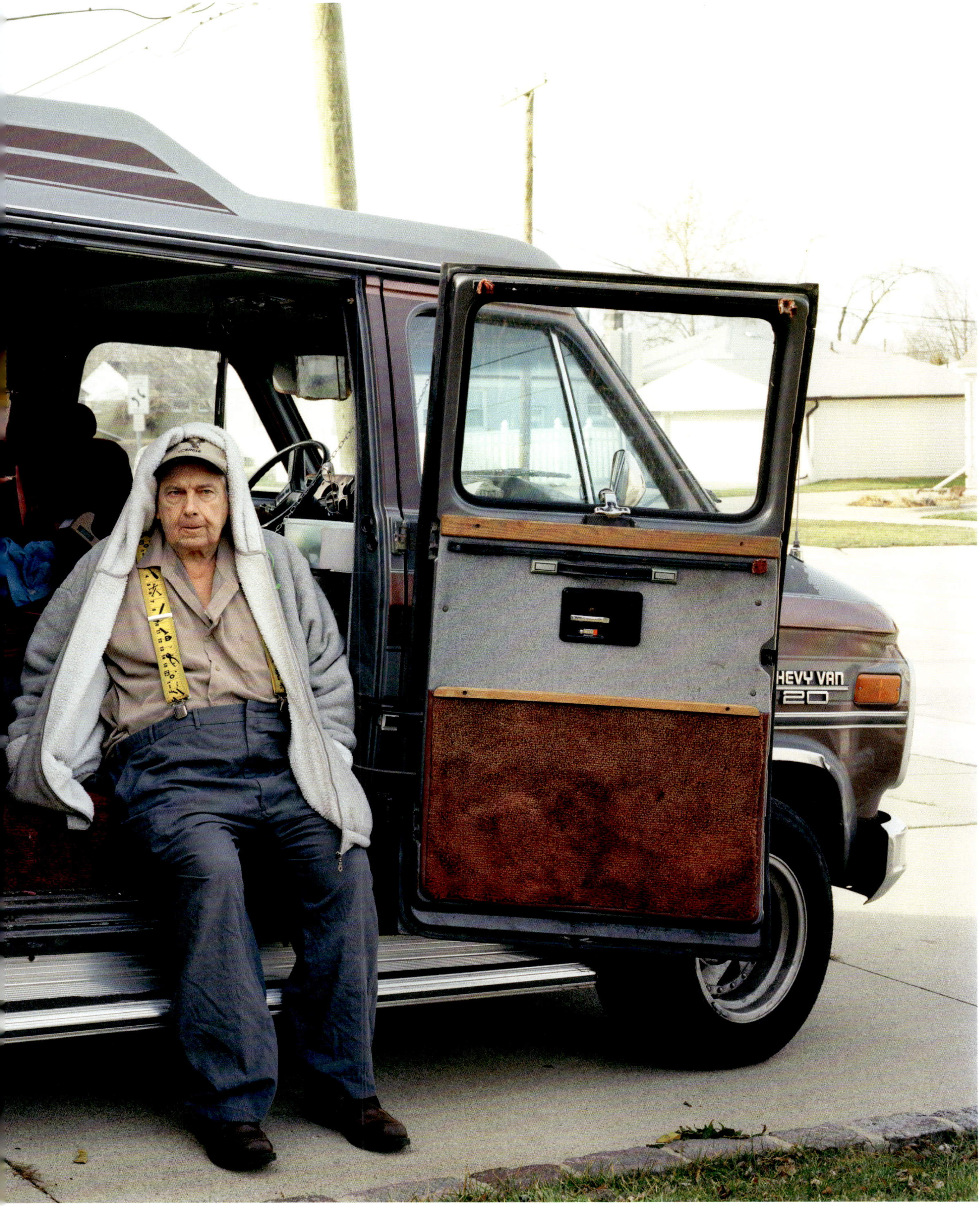
HEVY VAN
20

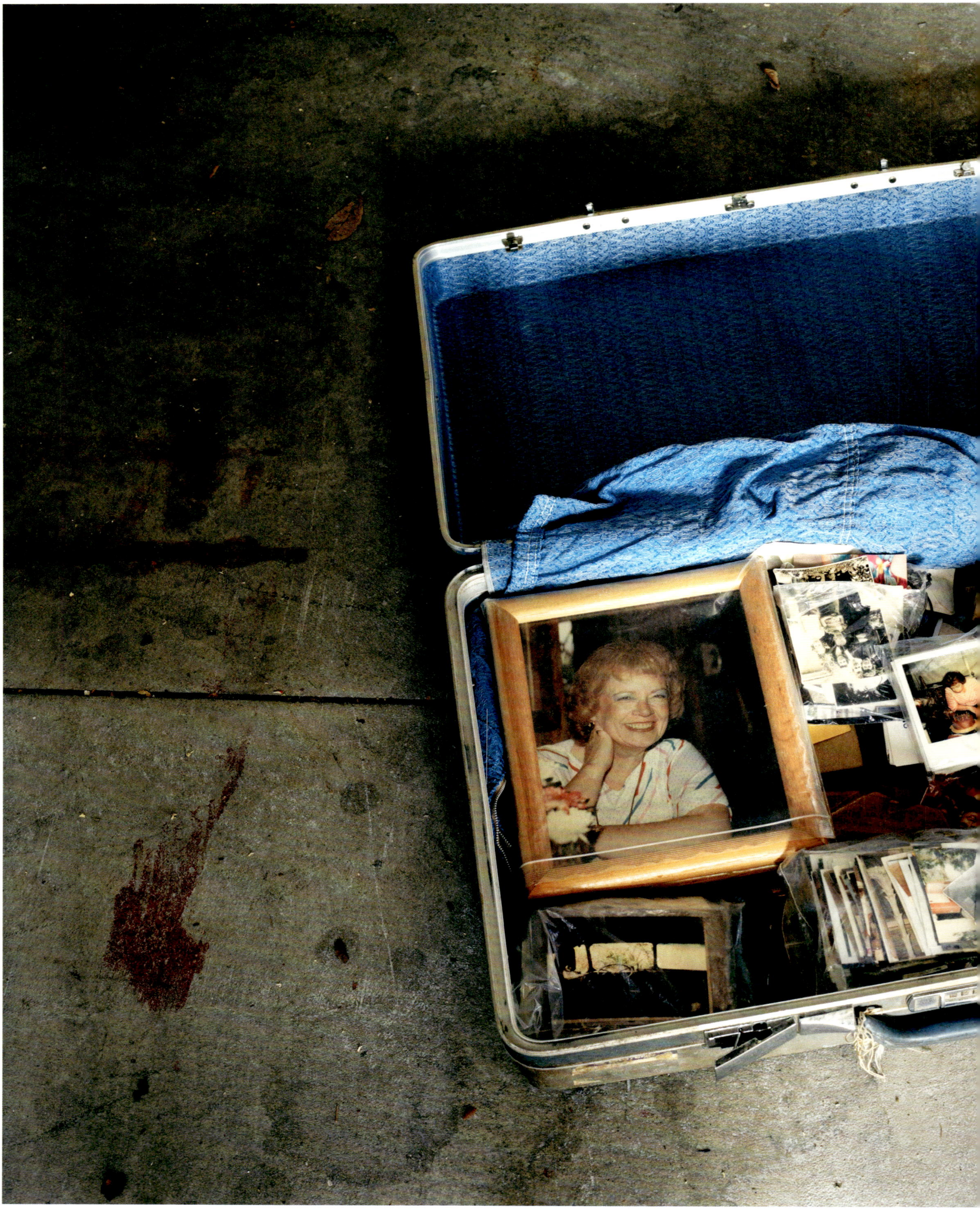

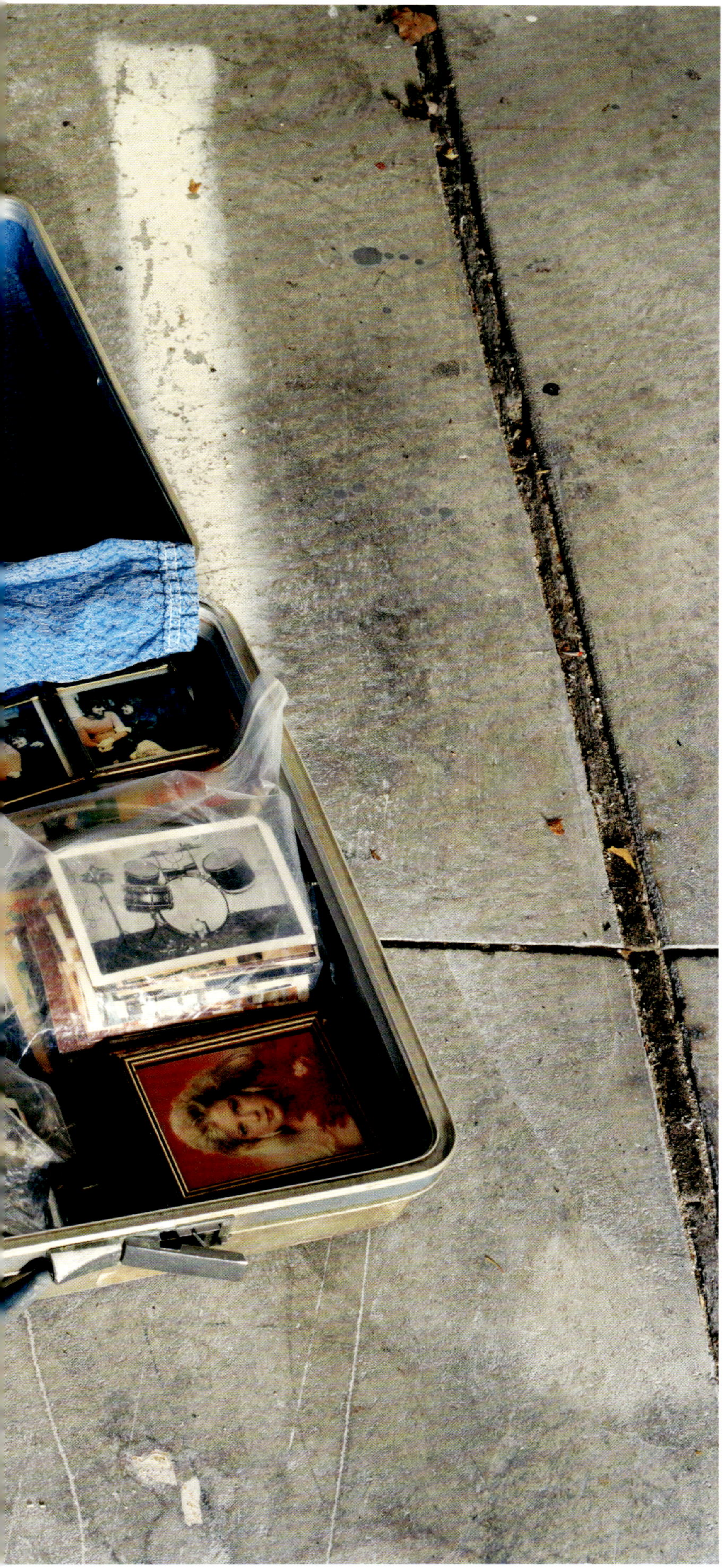

GRANDPA BREDO

NEDERLAND, COLORADO / OSLO

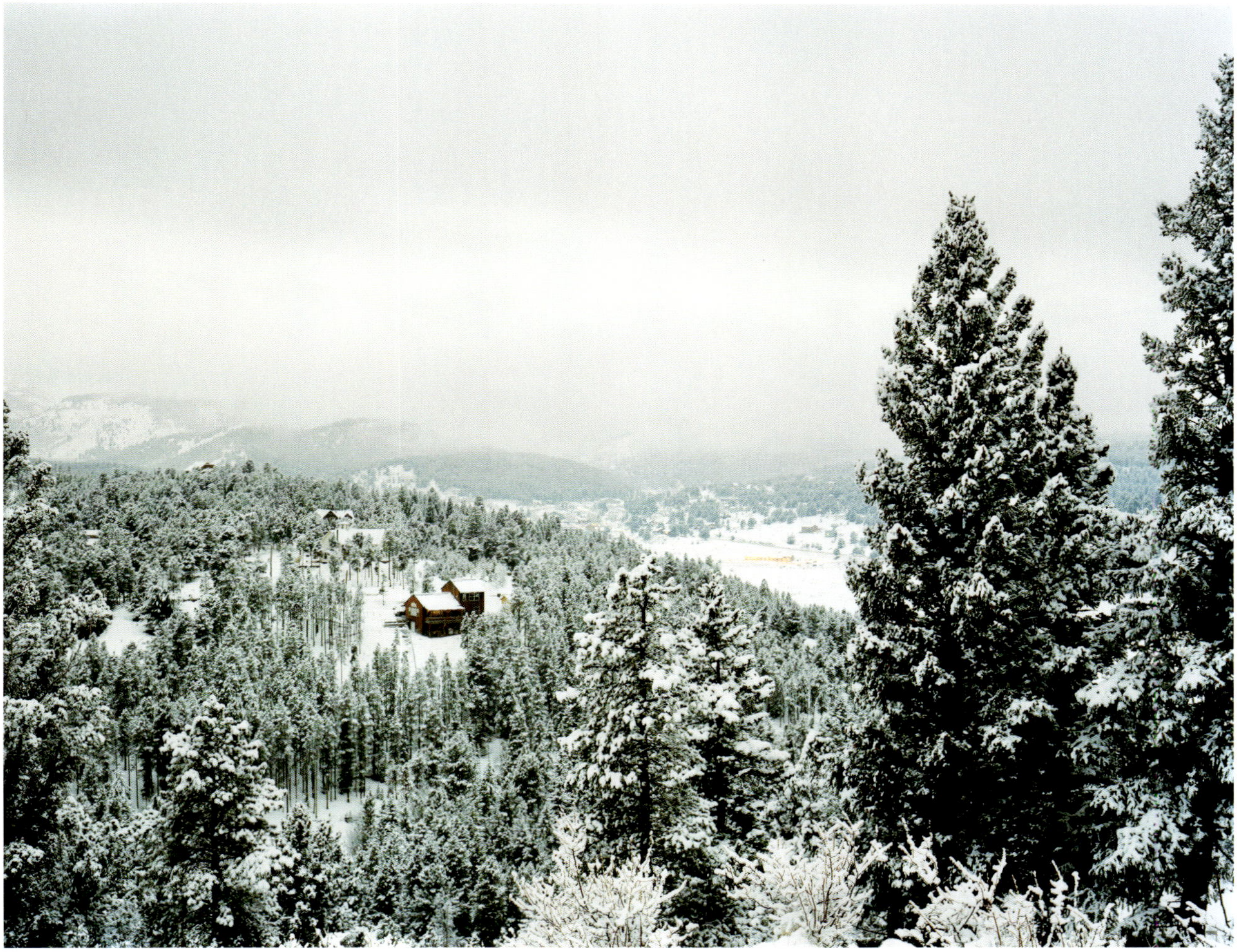

In 1980, Trygve Bauge, a Norwegian national, moved to the United States to pursue his interest in life-extension. Believing in individual liberty and the right to travel freely, he didn't apply for a visa.

In the autumn of 1991, Trygve and his mother, Aud, purchased a plot of land above Nederland, Colorado, and started building the Rocky Mountain Cryonics Facility — a family home with a connected blast shelter and cryonics laboratory. The entire construction was designed to be earthquake proof, fireproof, stormproof and to withstand a thermonuclear war.

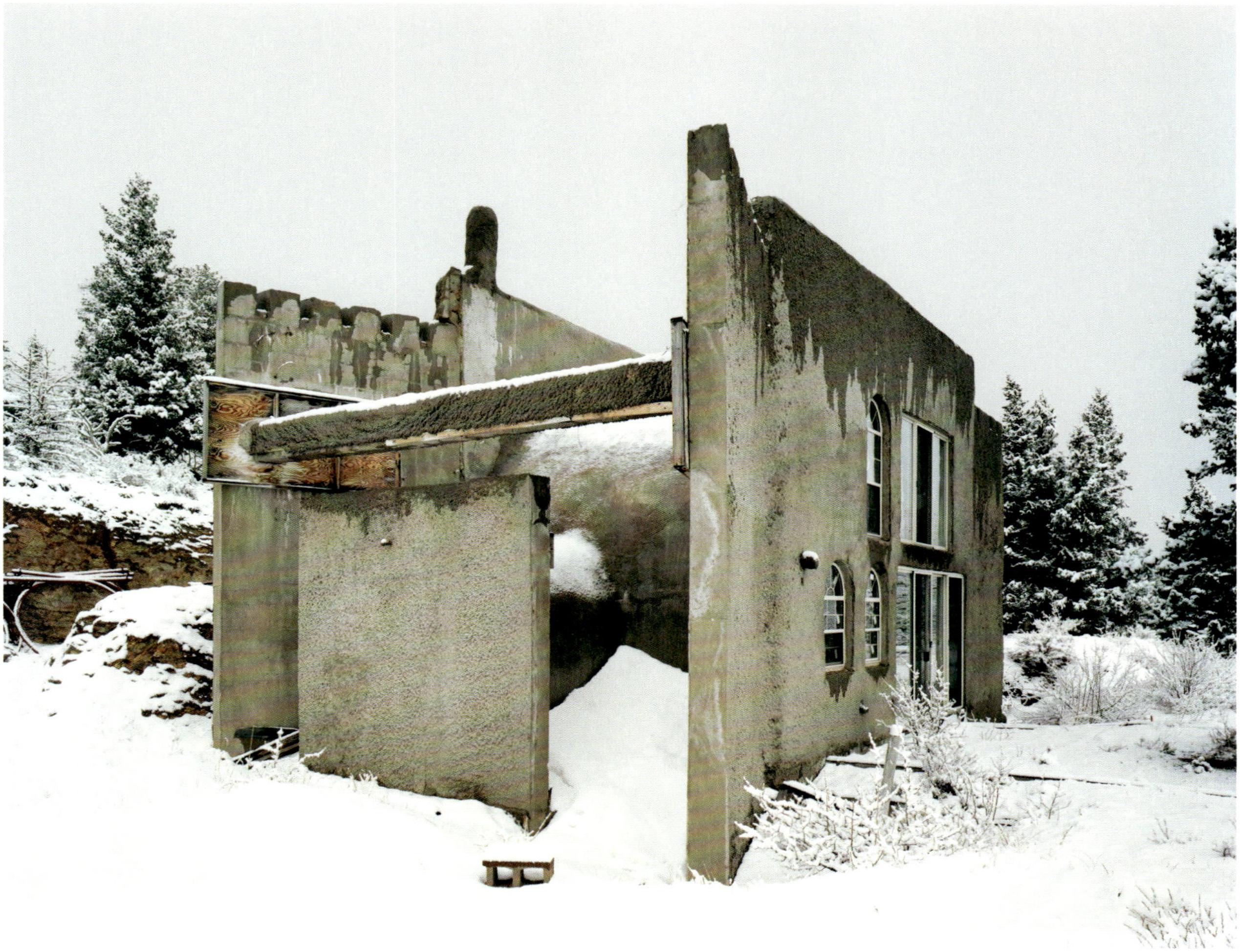

When his grandfather, Bredo Morstoel, died in Norway, he arranged for his body to be shipped to the USA for cryopreservation. Because the facility was not finished, Trygve temporarily stored him in a wooden shed.

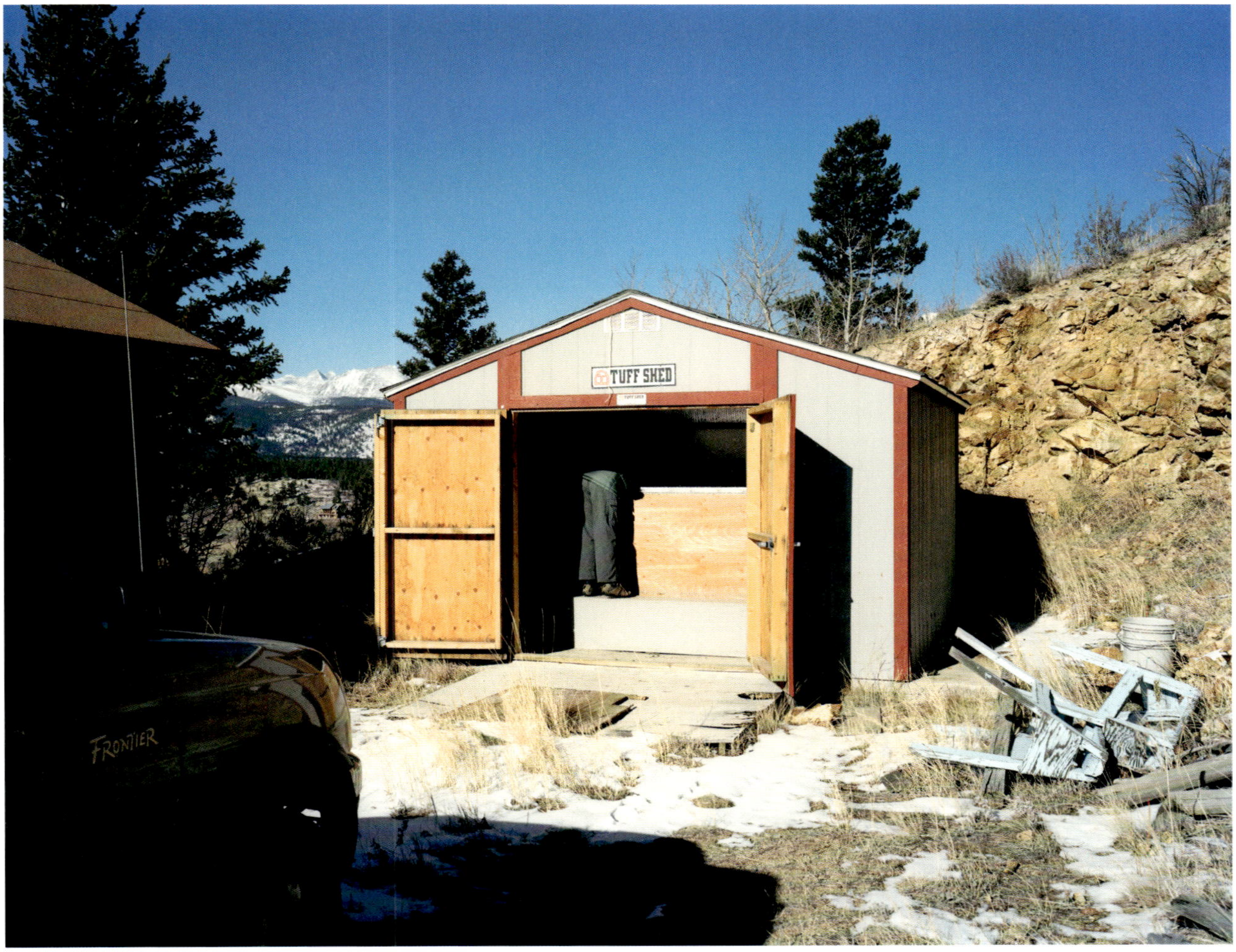
TUFF SHED
FRONTIER

Bredo was kept inside an insulated box of dry ice. In May 1994, after living in the USA for 14 years as an undocumented alien, Trygve was deported and forced to return to Norway. In February 1995, Aud was also deported.

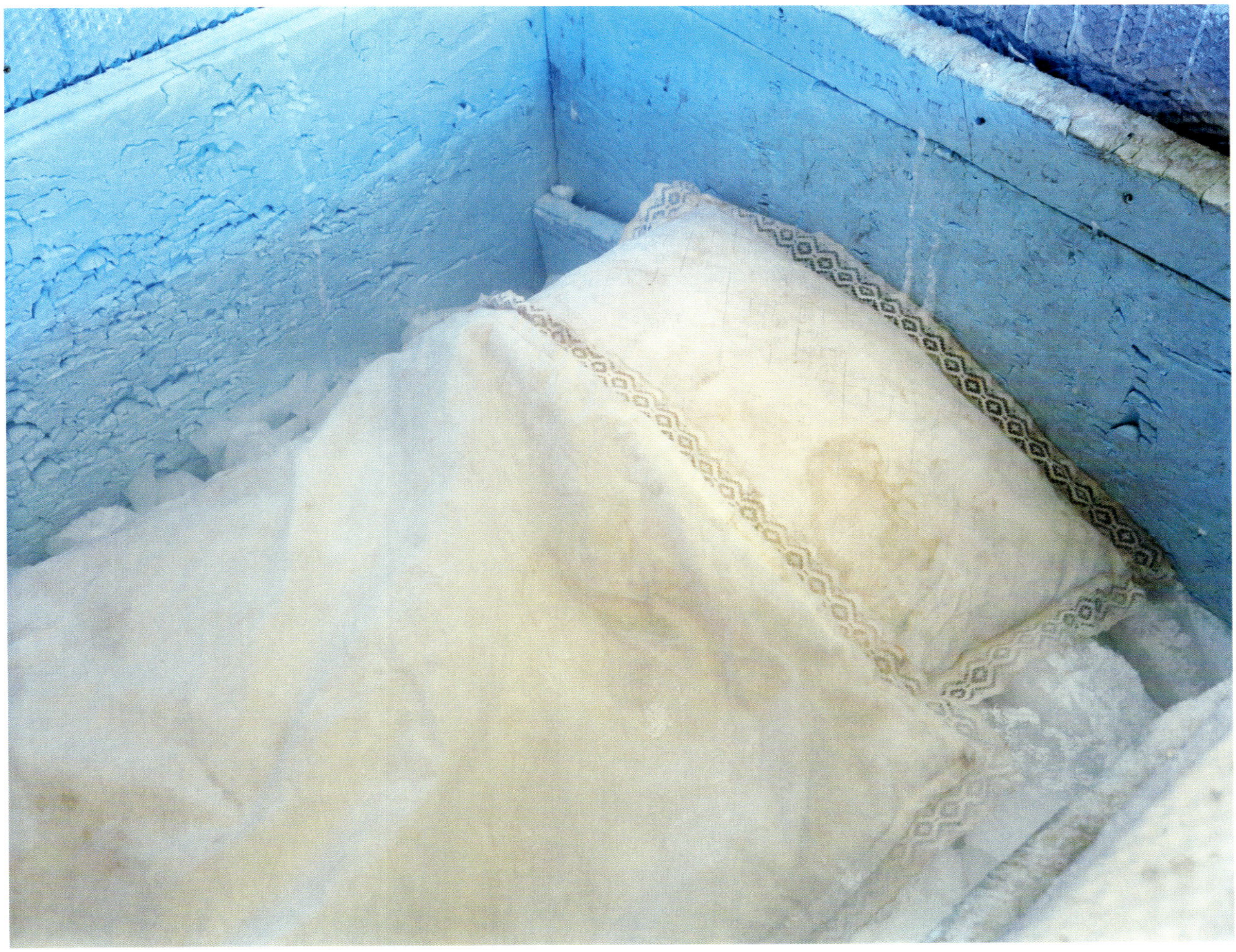

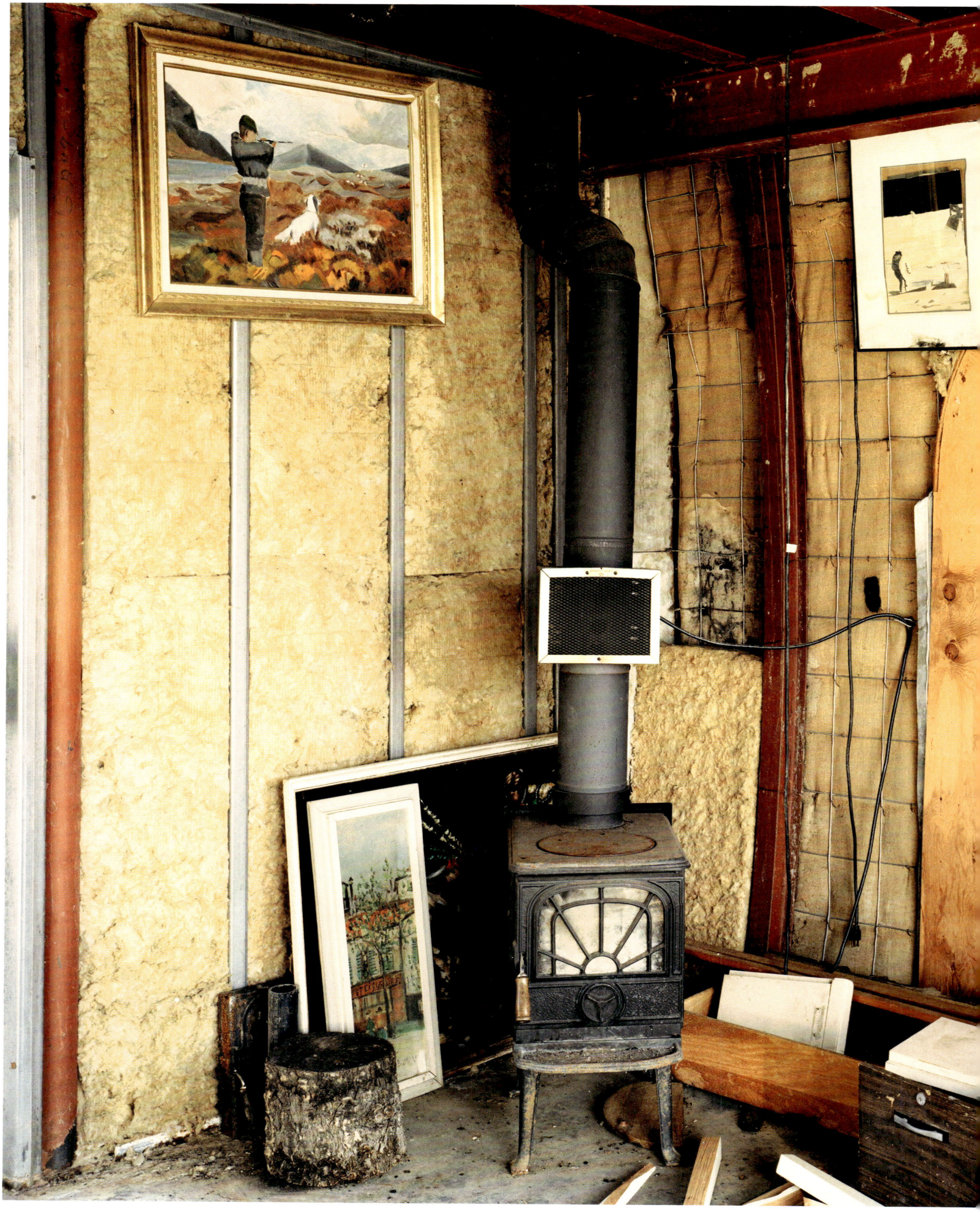

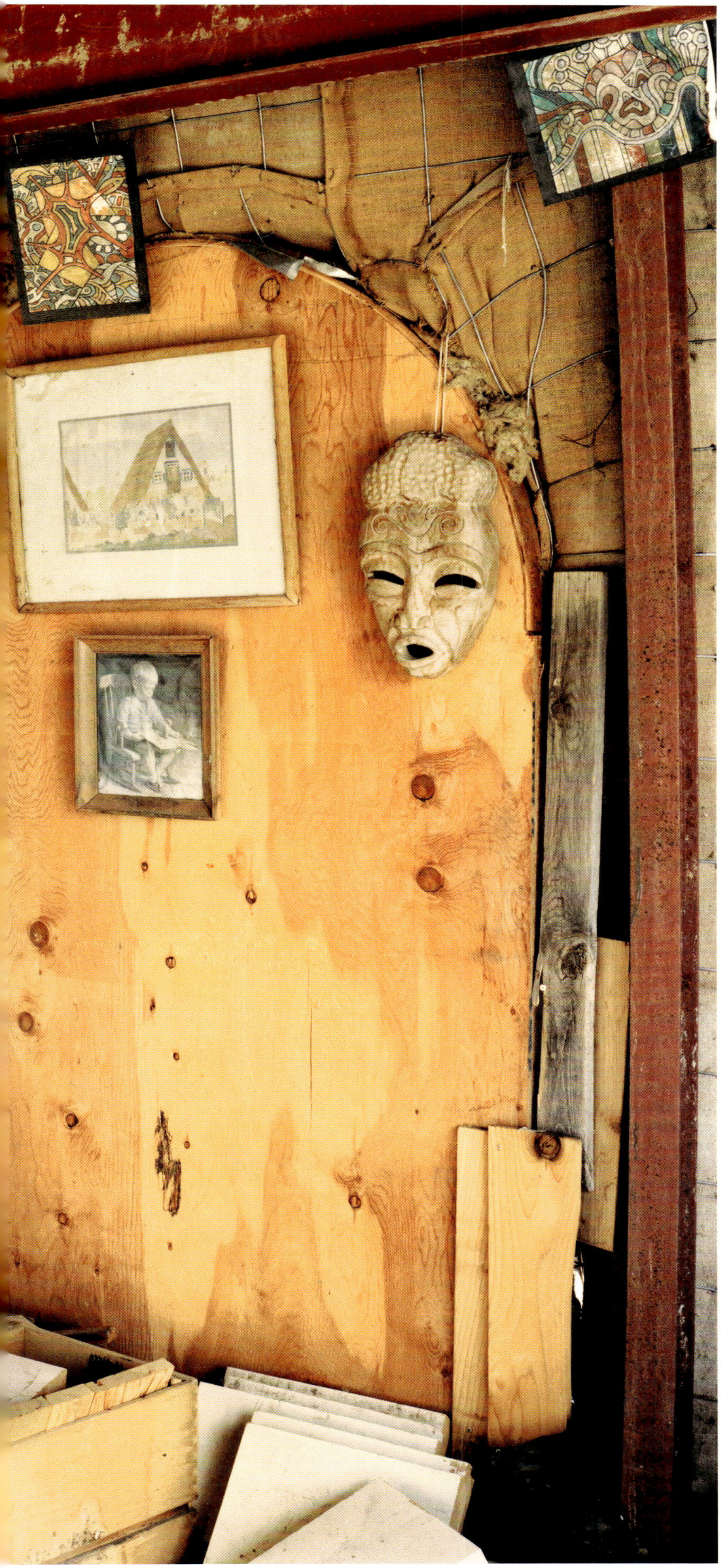

From 1994, Aud shared a flat with Trygve in the suburbs of Oslo, until she began to suffer from schizophrenic episodes. She is now staying at a local hospital for treatment. Trygve continues to pursue his interest in life-extension and plans, one day, to publish his 'Joyful Life-Extension Manual'. He is also a passionate ice-bather. He once stayed in the water for one hour and four minutes.

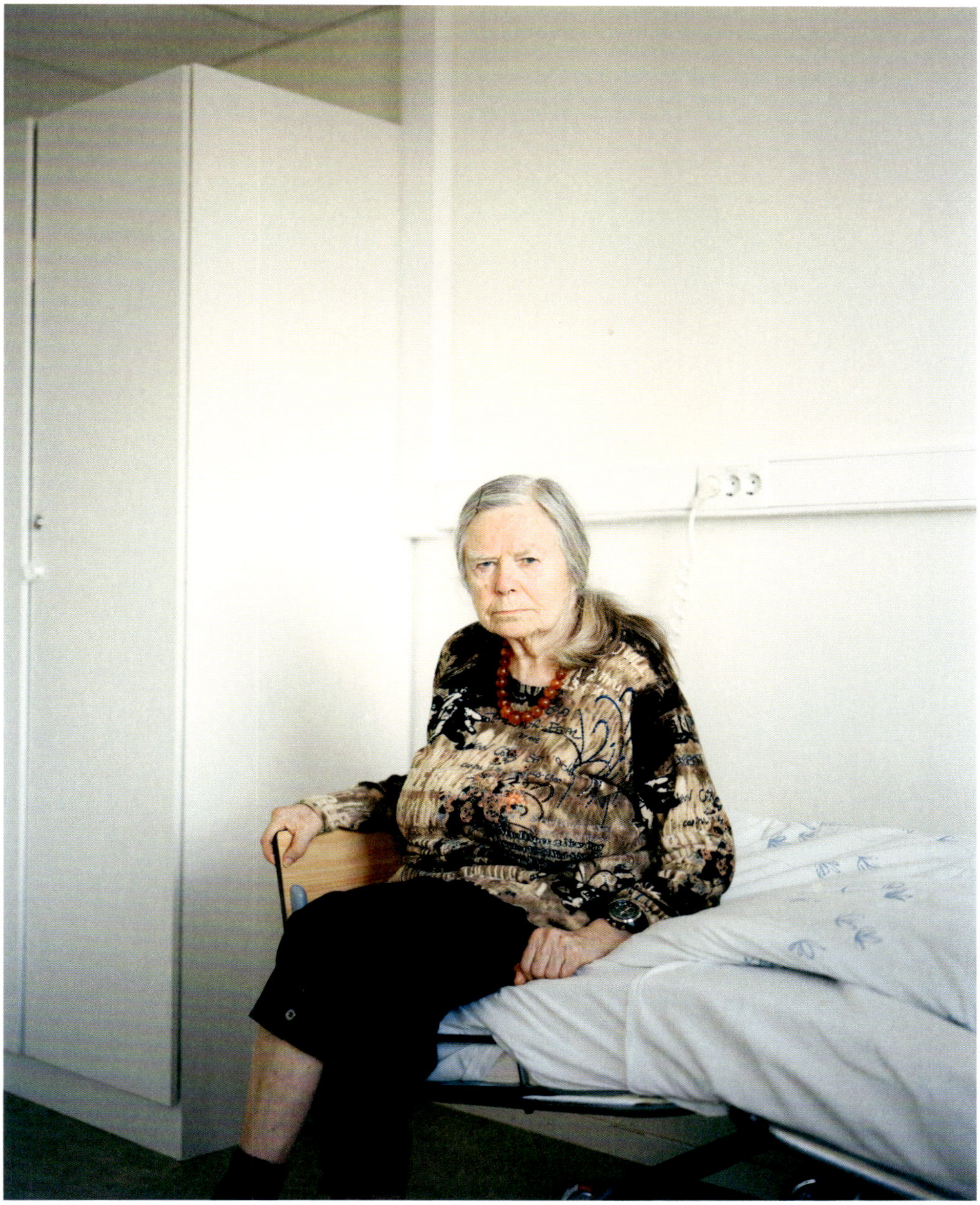

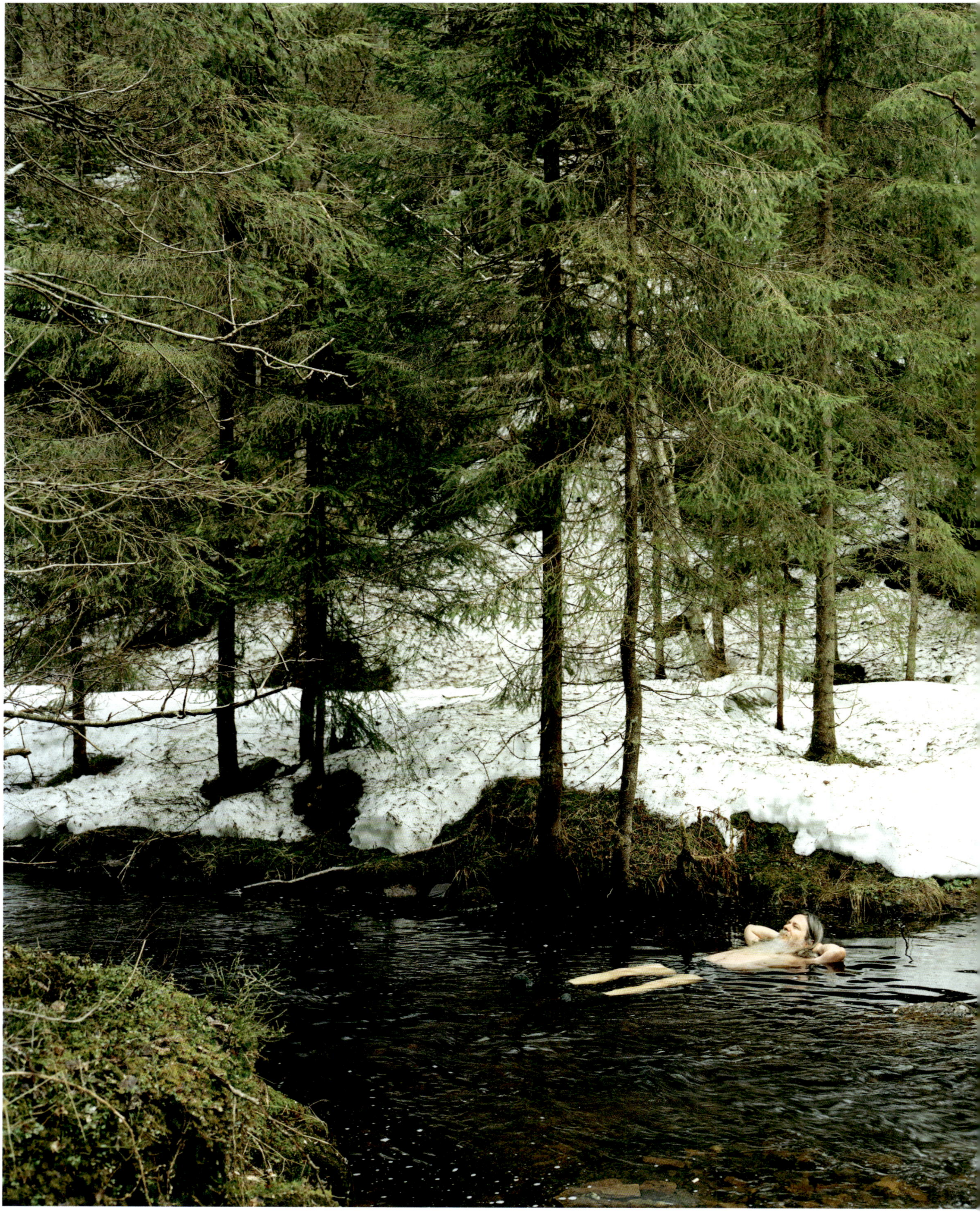

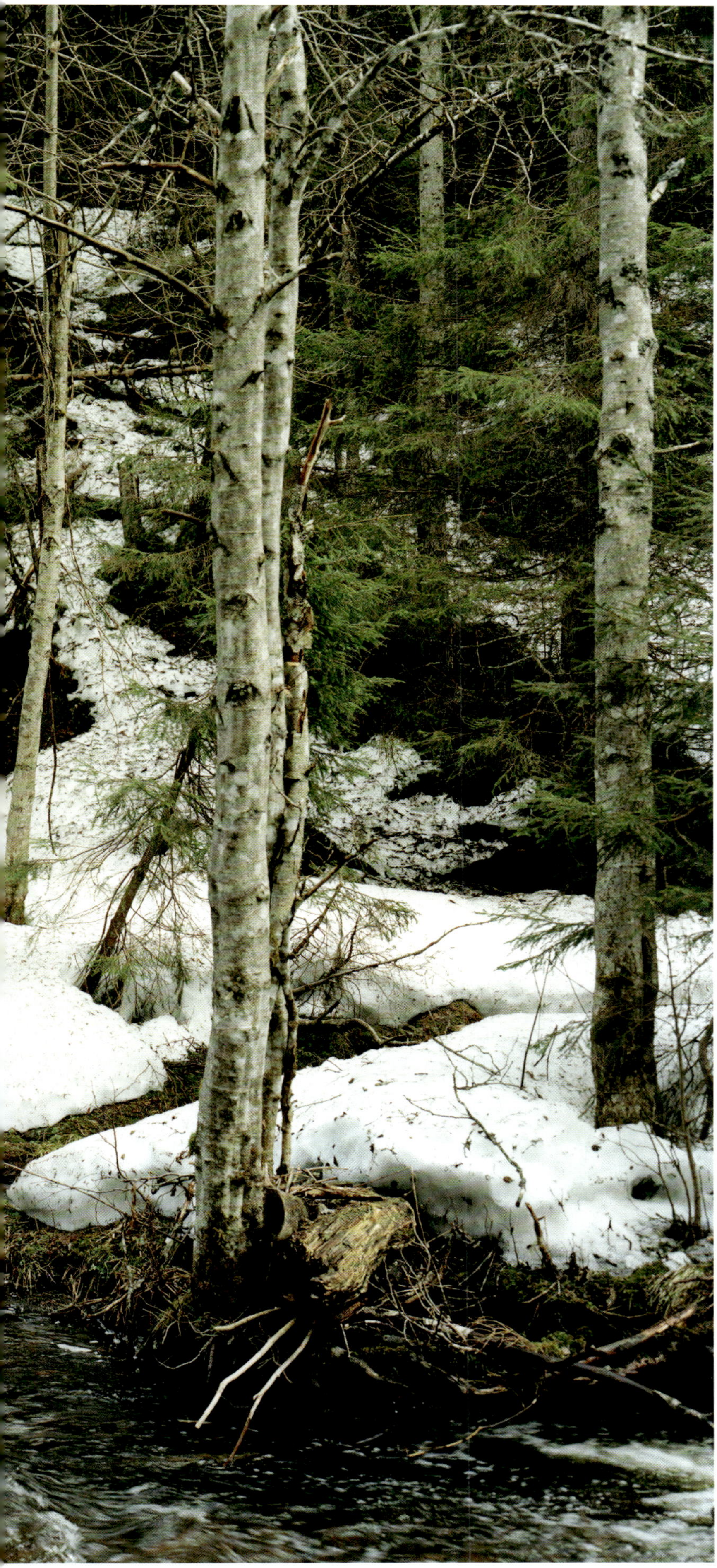

Grandpa Bredo is still frozen in Nederland, Colorado, and is maintained by a local caretaker. The caretaker replenishes Bredo's dry ice every two weeks. The monthly costs to Trygve and Aud is $1100.

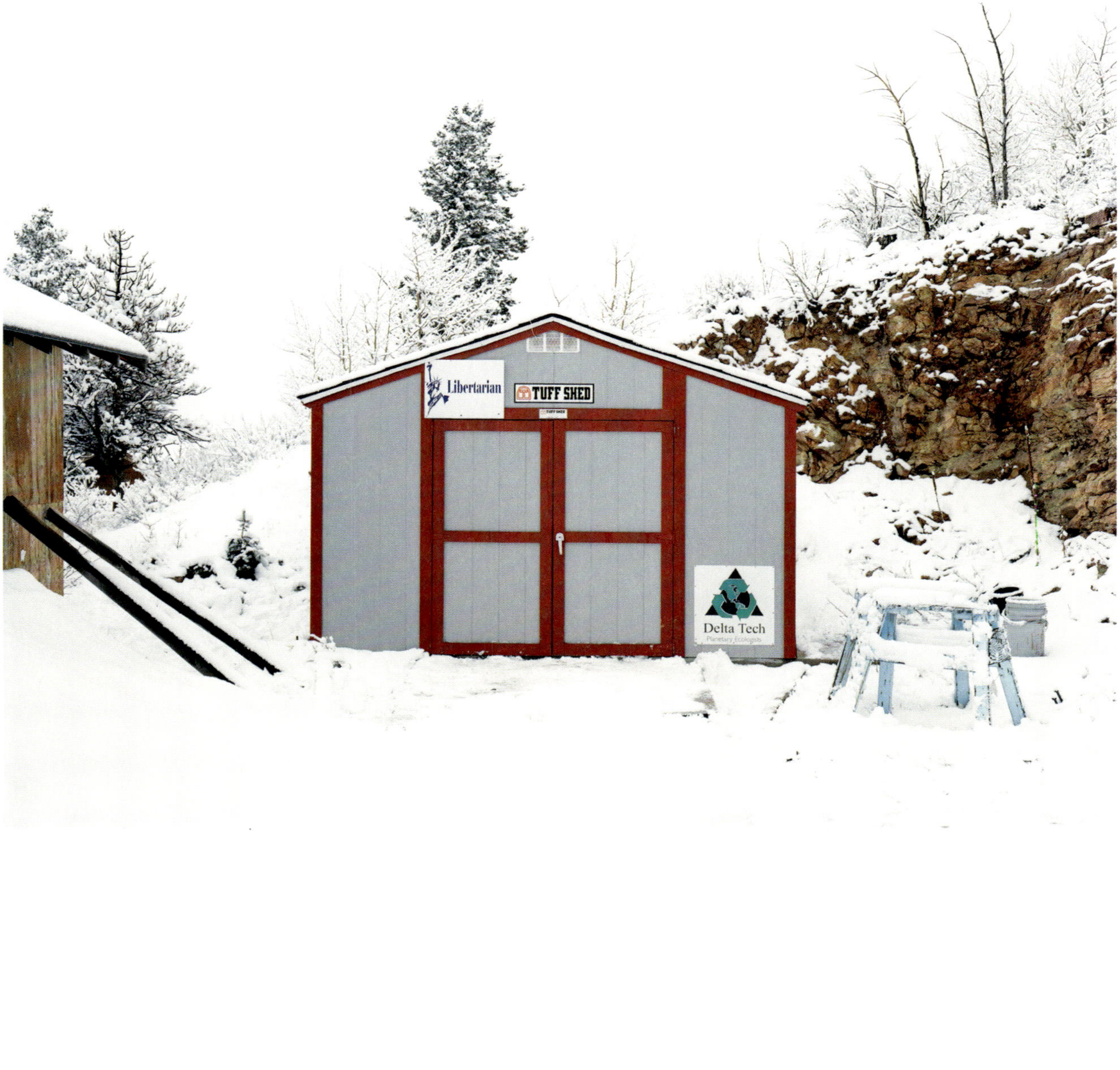
Libertarian
TUFF SHED
Delta Tech
Planetary Ecologists

KRIORUS PART 2

ALABUSHEVO, MOSCOW

The most precious thing for a person is life. People are always dreaming of an eternal life and these dreams of immortality gave rise to different fantastic religions, beliefs and faiths. My mama knew the meaning of life and, from a young age, she dreamt of a real and fantastic immortality. She became a scientist and she proved to me, scientifically, that a personality could be saved and restored in the future. Cryonics is one of the most realistic technologies offering us such an opportunity. My mama taught me that the meaning of life is to understand and to study the world and improve it. So this understanding of life makes immortality philosophically reasonable and justifiable.

I love my mama most of all, more than I love myself, because she loved me and all the people more than she loved herself. That's why I decided to cryopreserve her and to keep her brain for the future. I don't know any other scientific alternative to a rotting body in a grave.

I'm happy that I managed to do the preservation in time. Mama will be one of the first people to experience a real immortality and who has managed to go into the future.

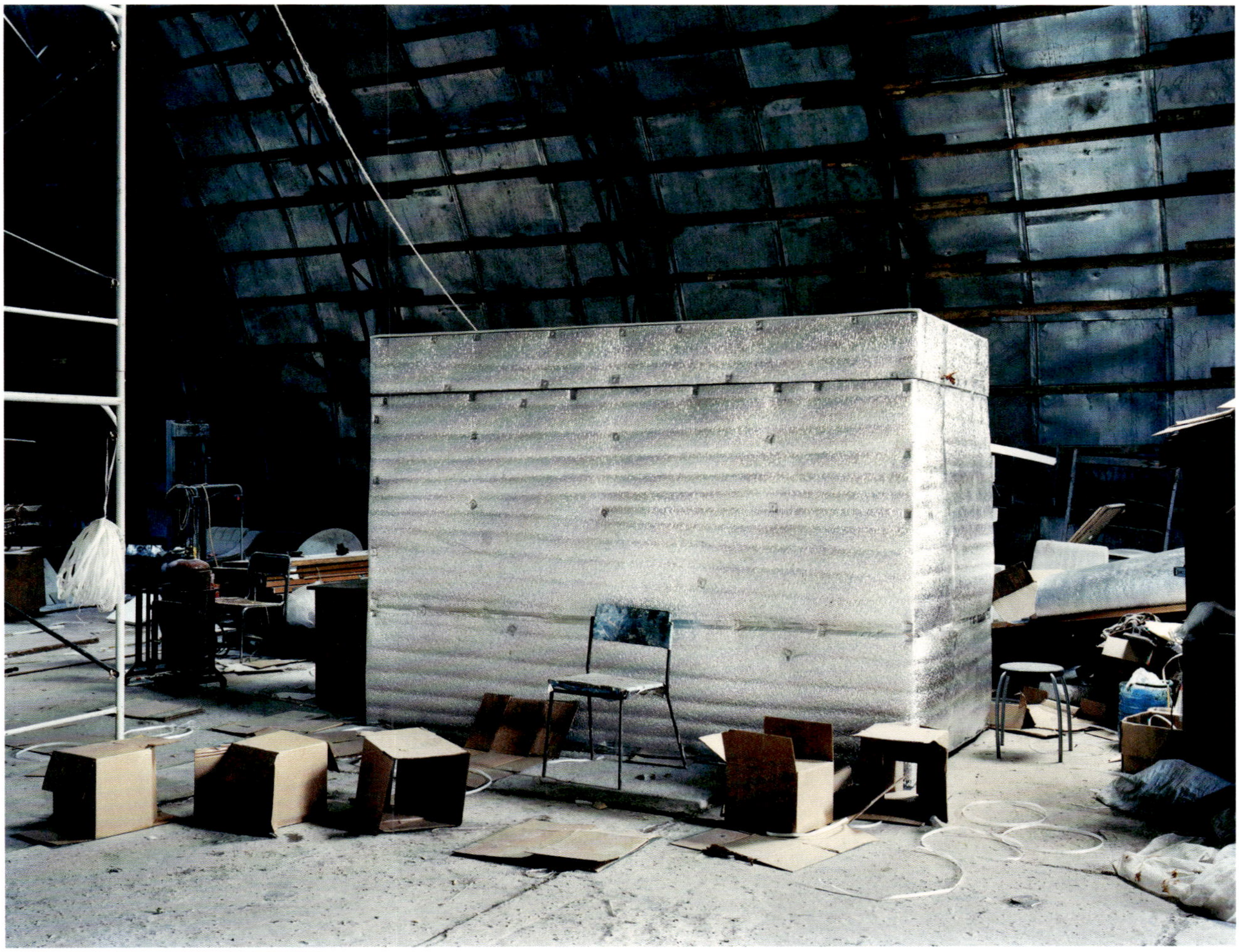

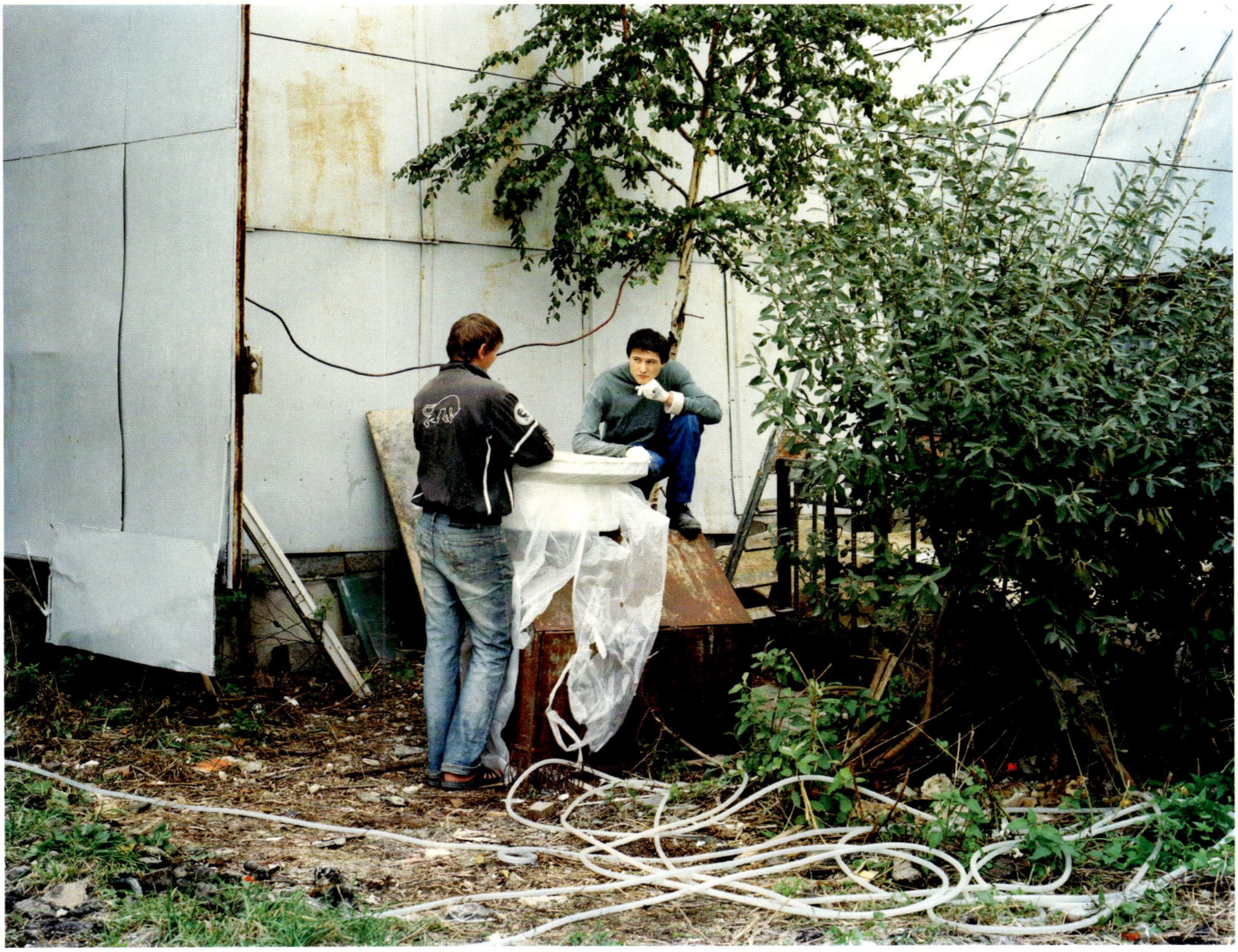

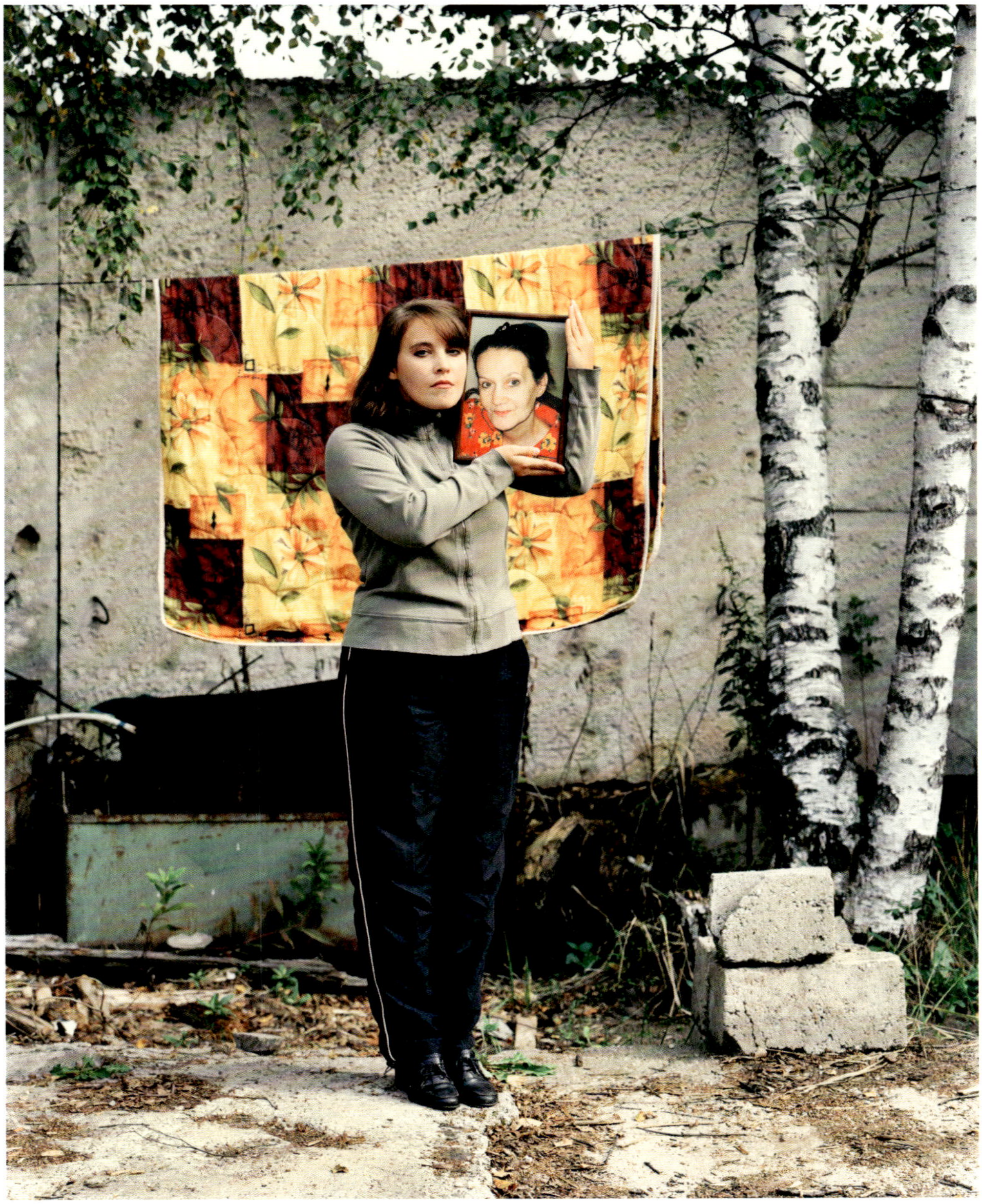

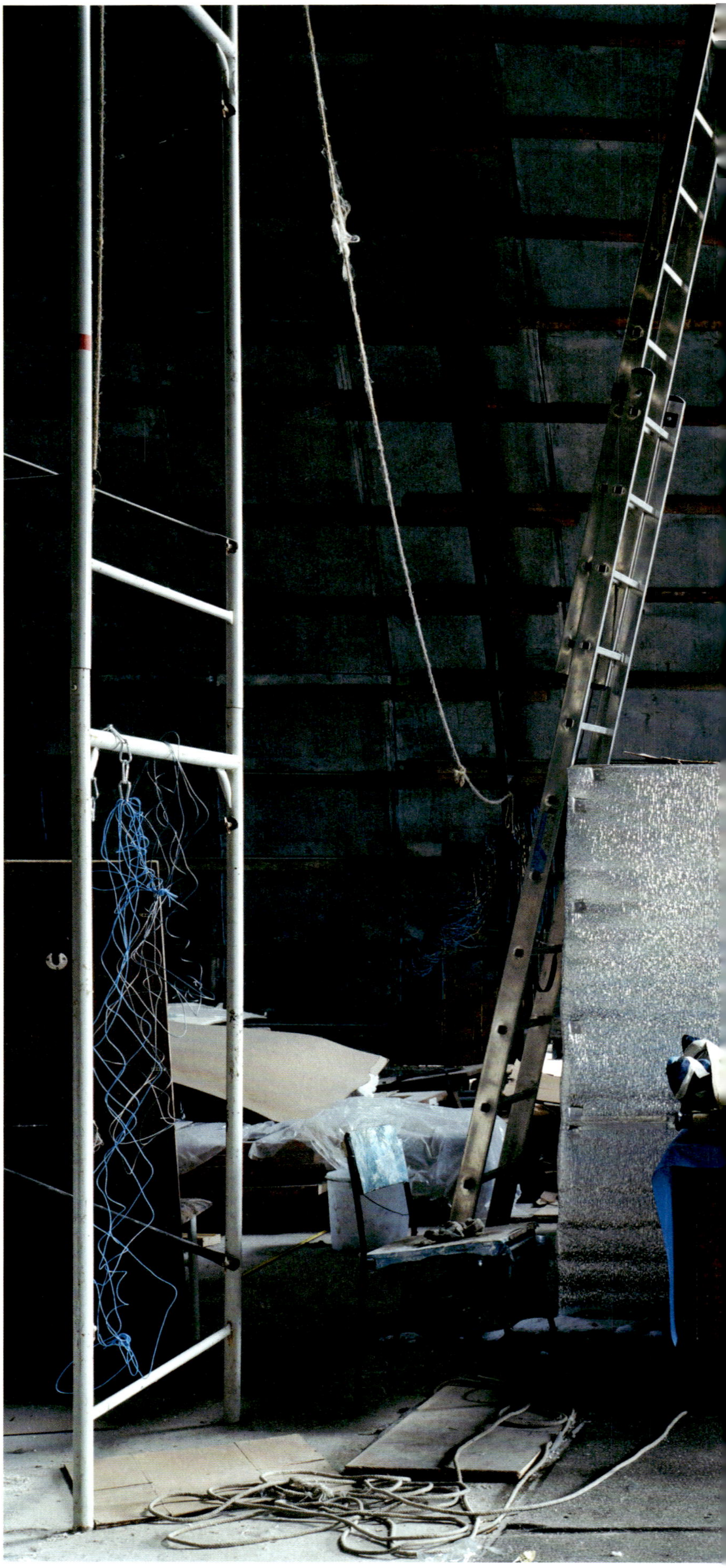

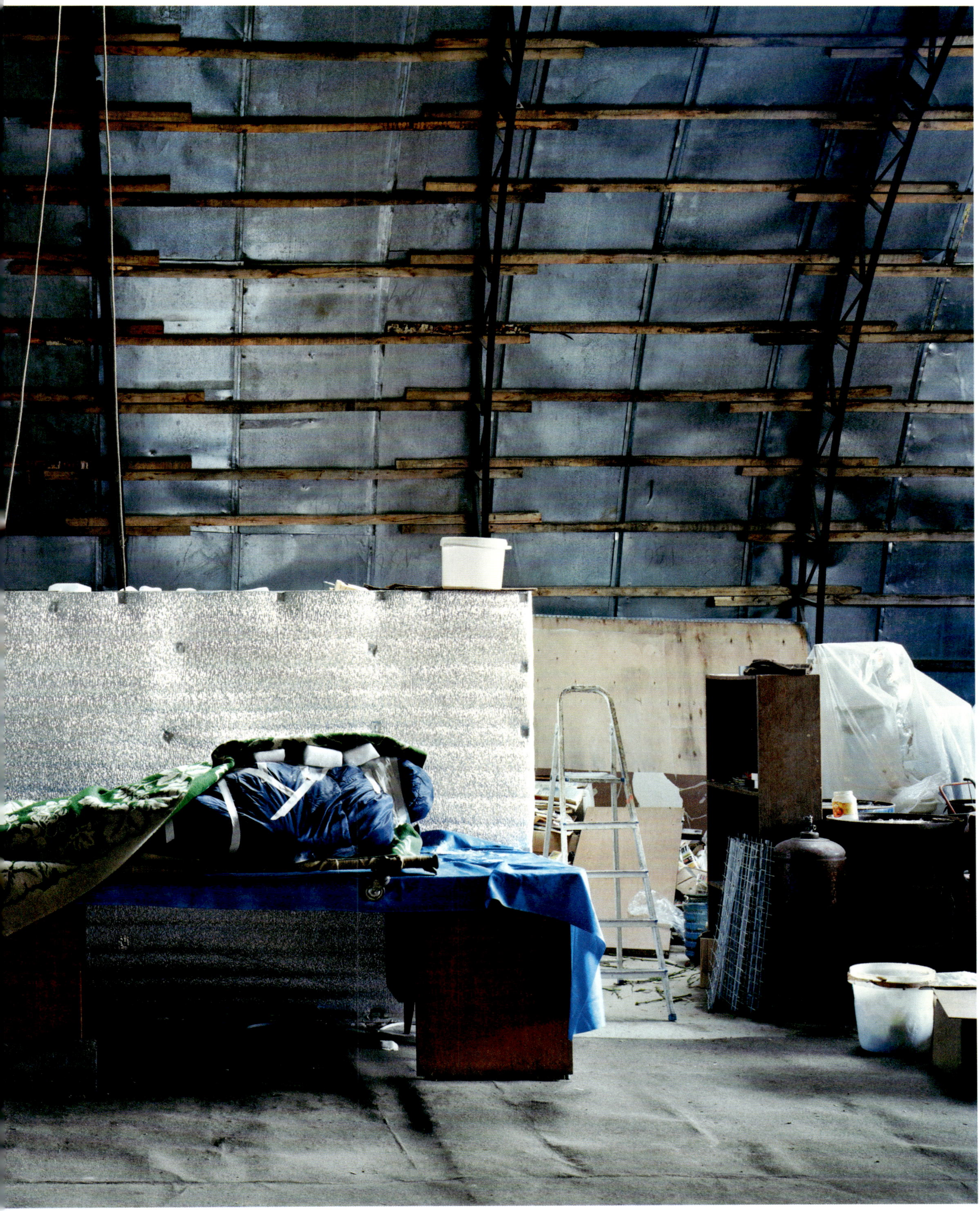

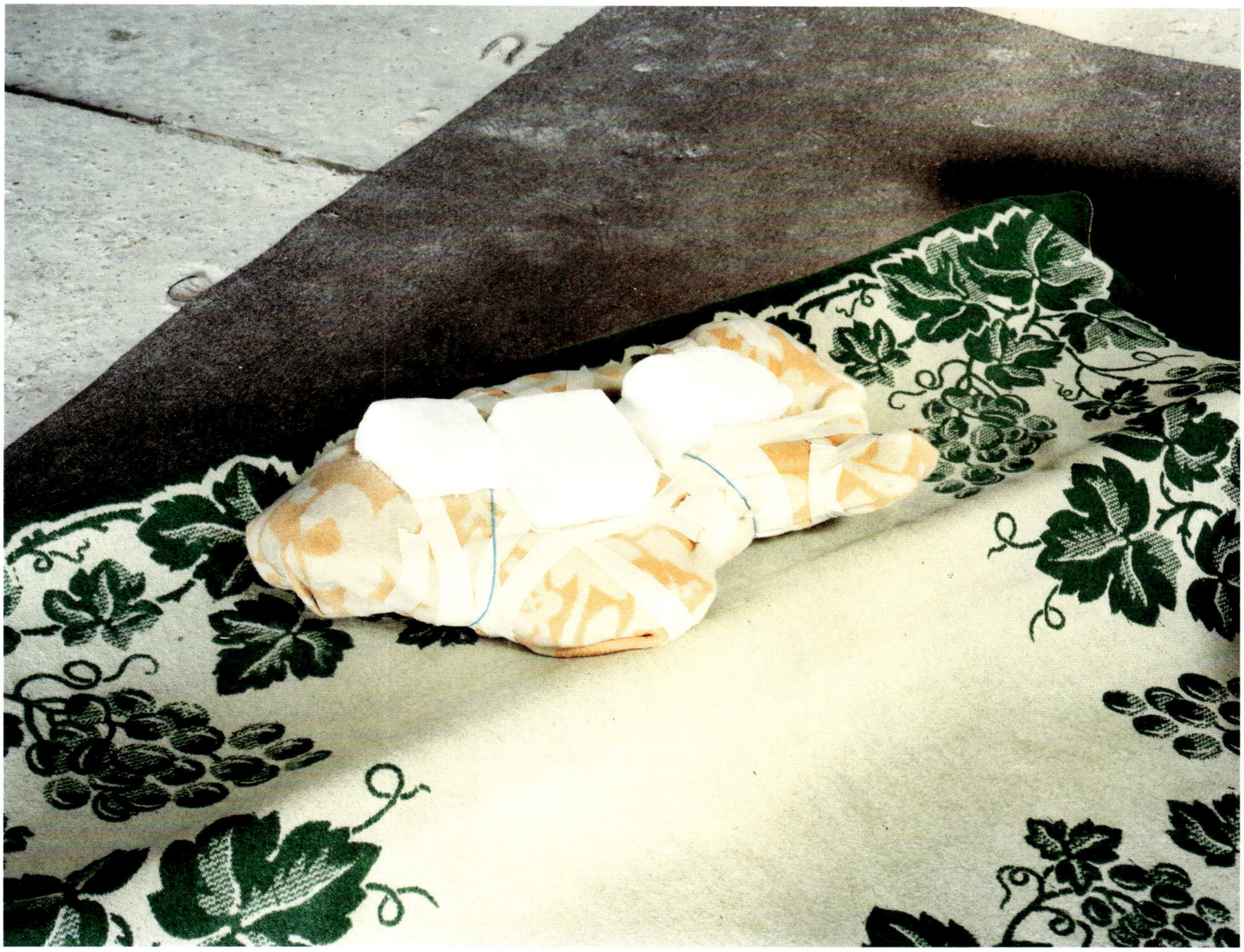

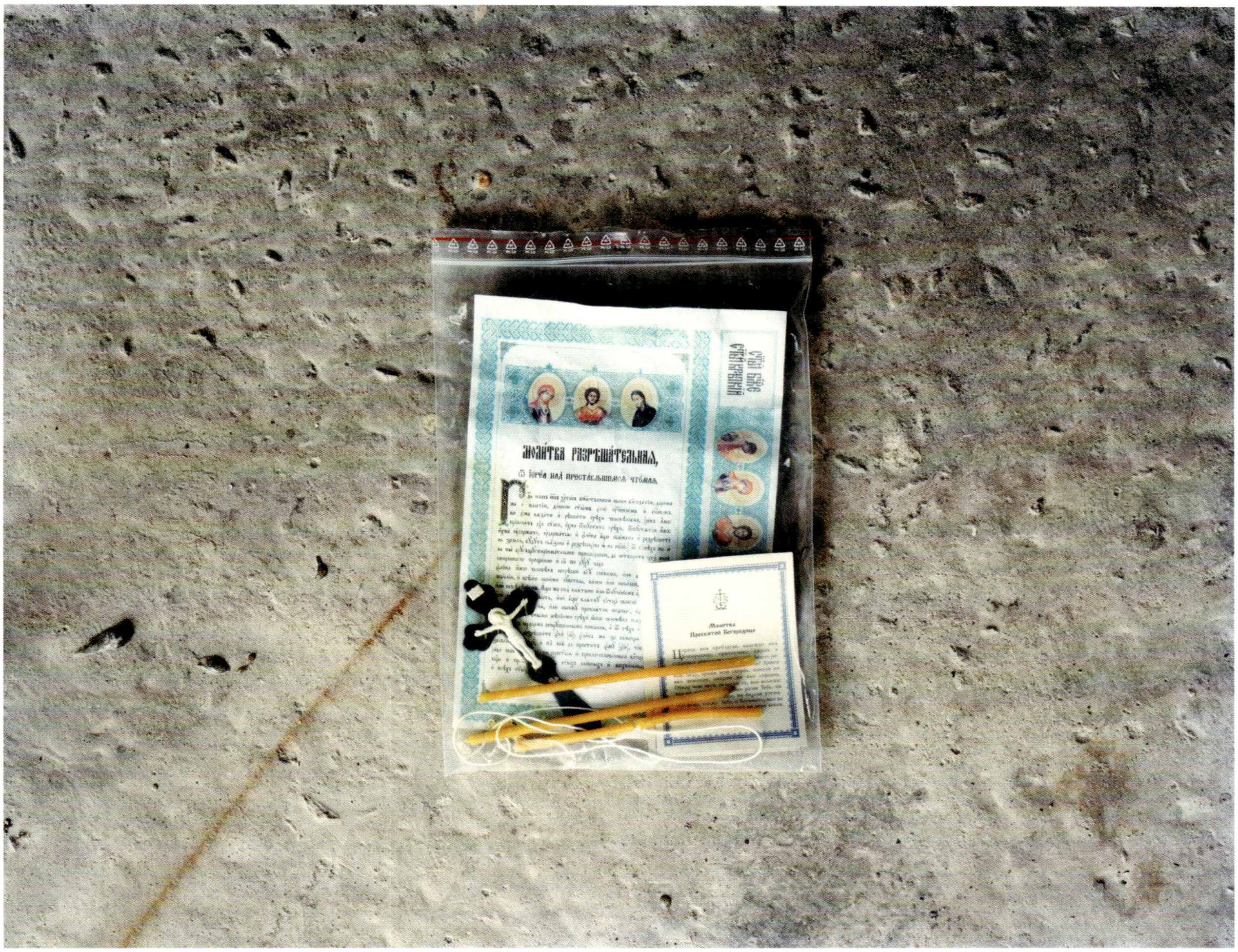
Молитва разрѣшительная,
Молитва
Пресвятой Богородице

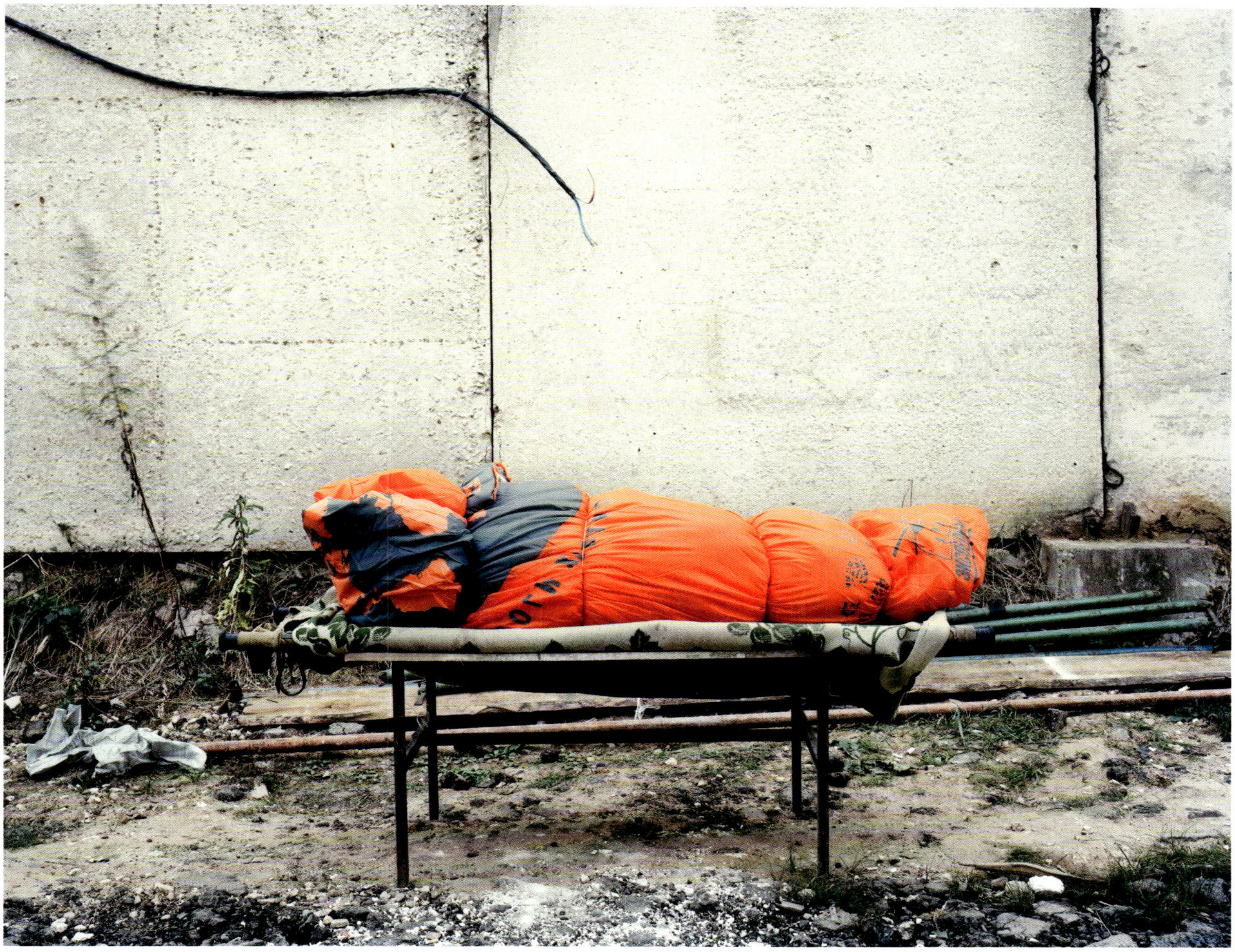

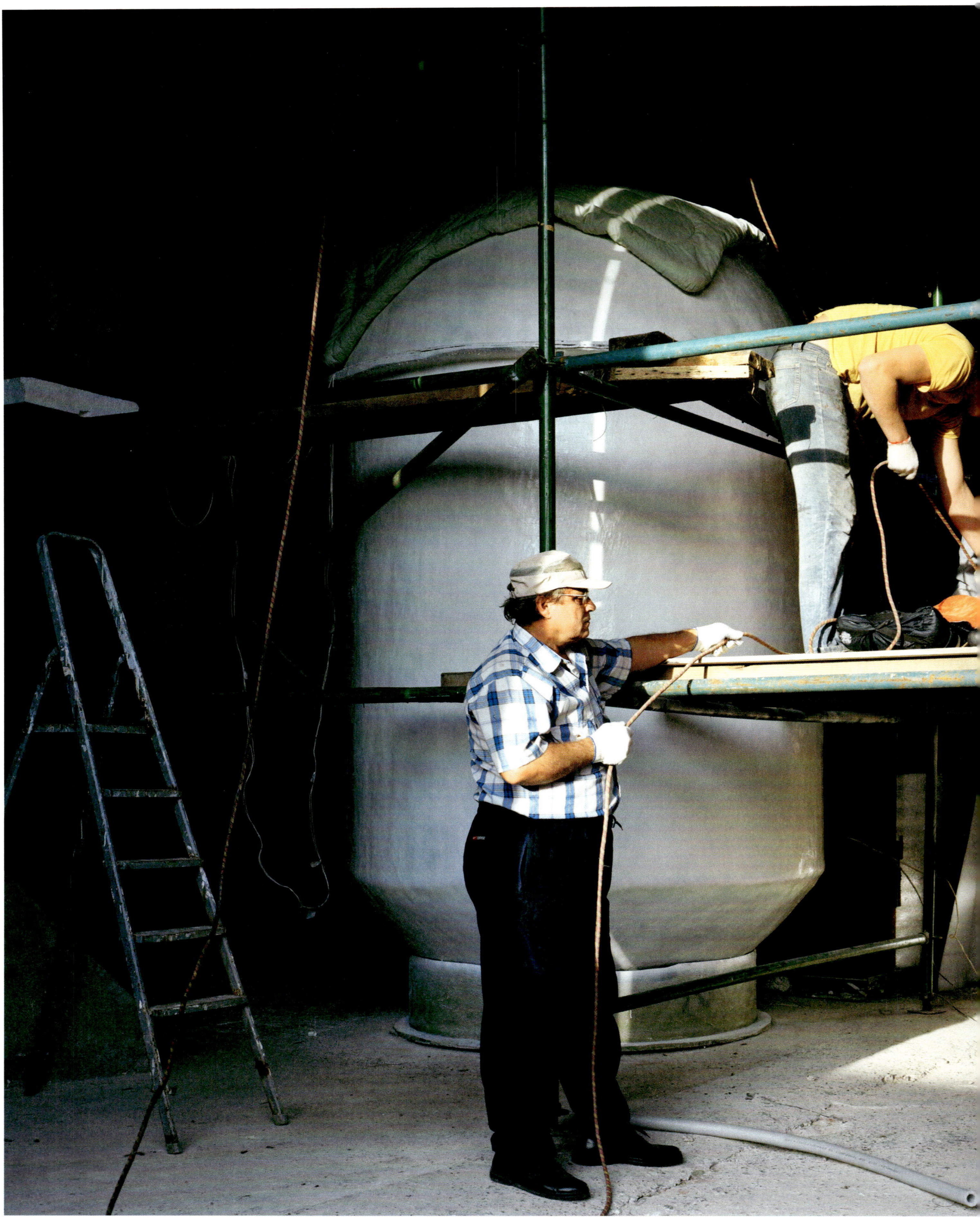

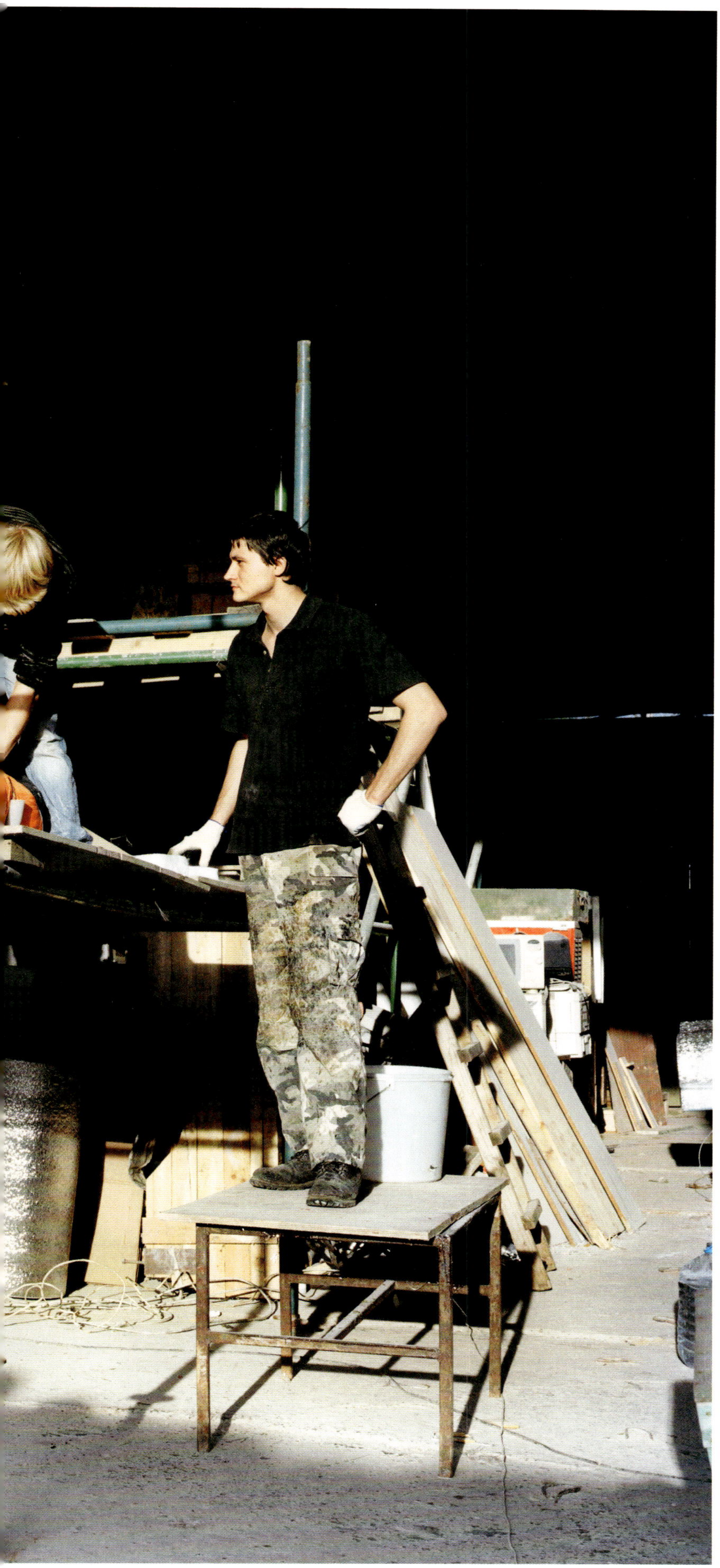

TIMESHIP

COMFORT, TEXAS

According to The Stasis Foundation, the Timeship project, near Comfort, Texas, will be a 'Fort Knox' for the cryopreservation of patients, organs, DNA, stem cells and tissue samples.

The Timeship is planned for a 800-acre swathe of land, which is bounded by the Guadalupe River and a winding stretch of country road.

Stasis's plans for the Timeship building are expected to resemble a medieval fortification covering six acres, but local planning and tax queries have delayed construction considerably.

Access to the site is strictly controlled by Stasis and detailed plans for the site remain a highly guarded secret.

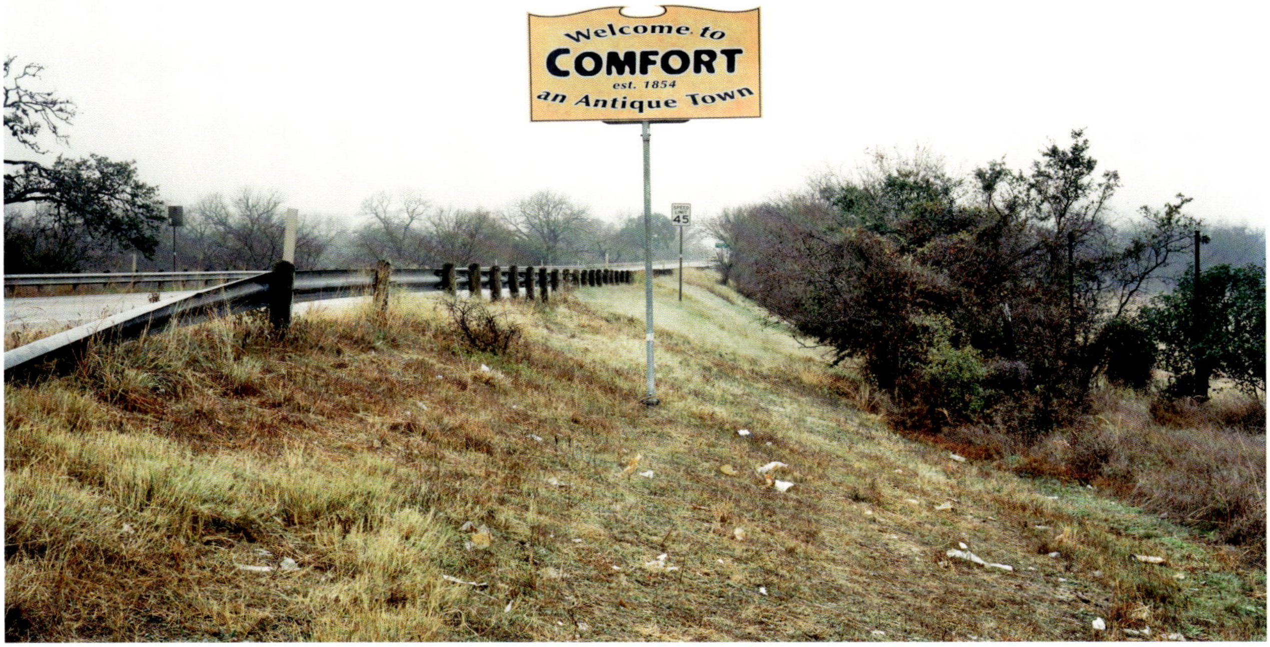
Welcome to
COMFORT
est. 1854
an Antique Town
45

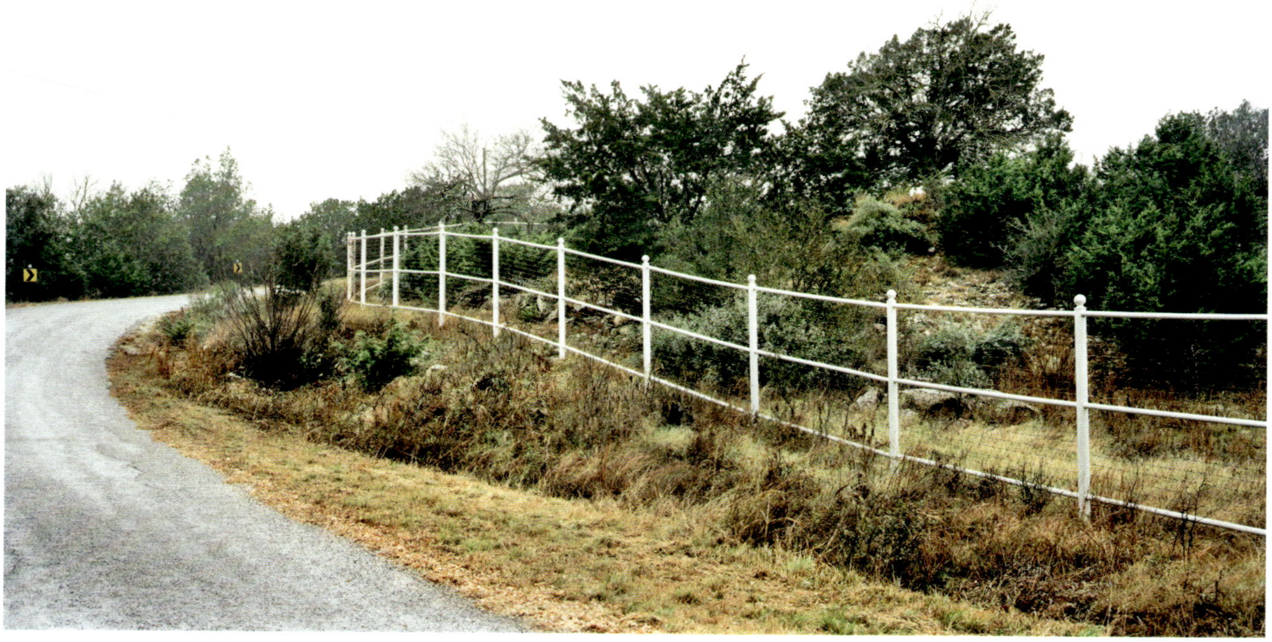

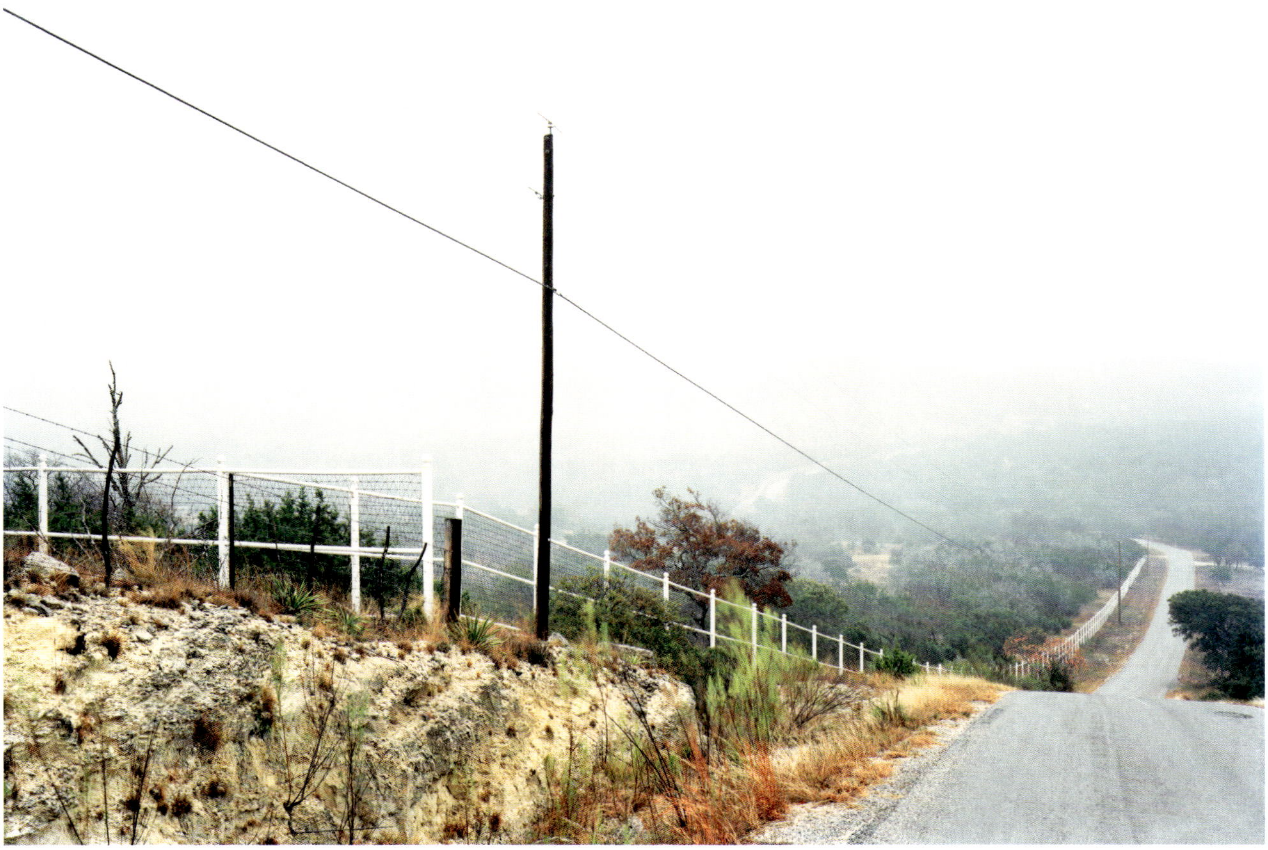

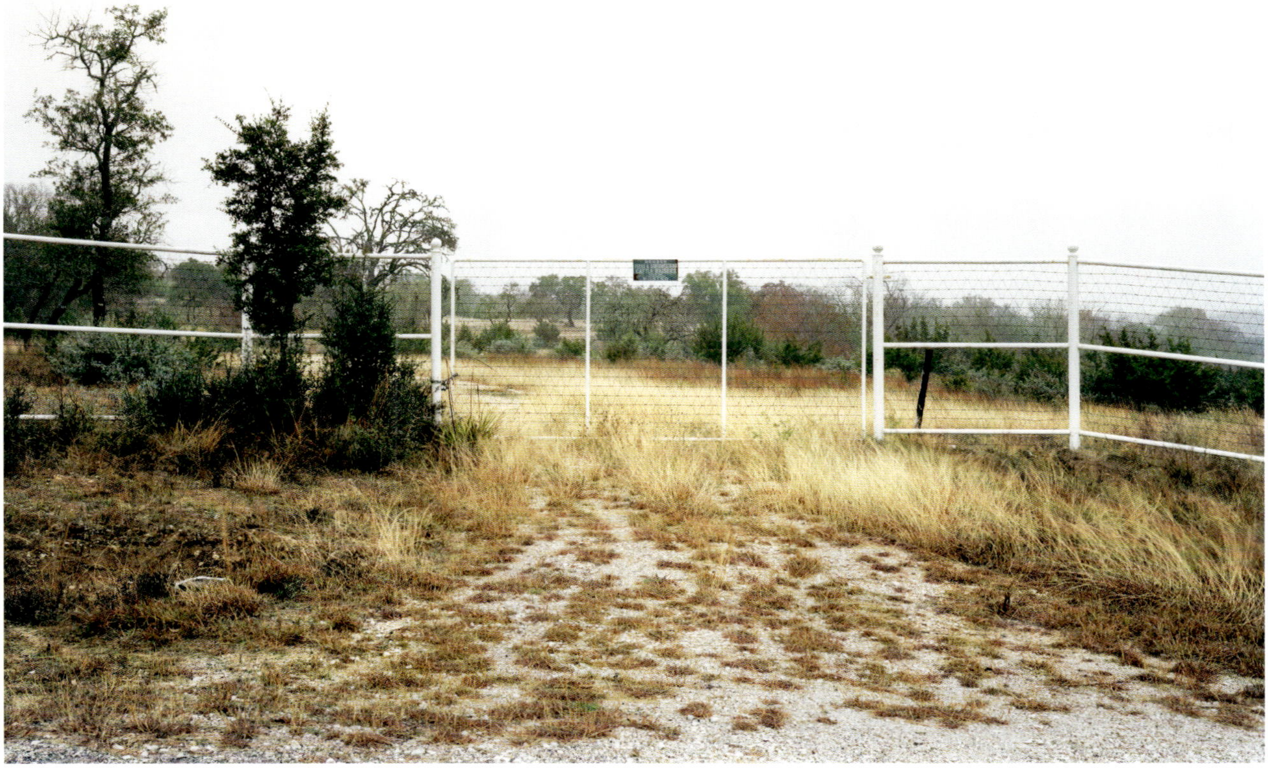

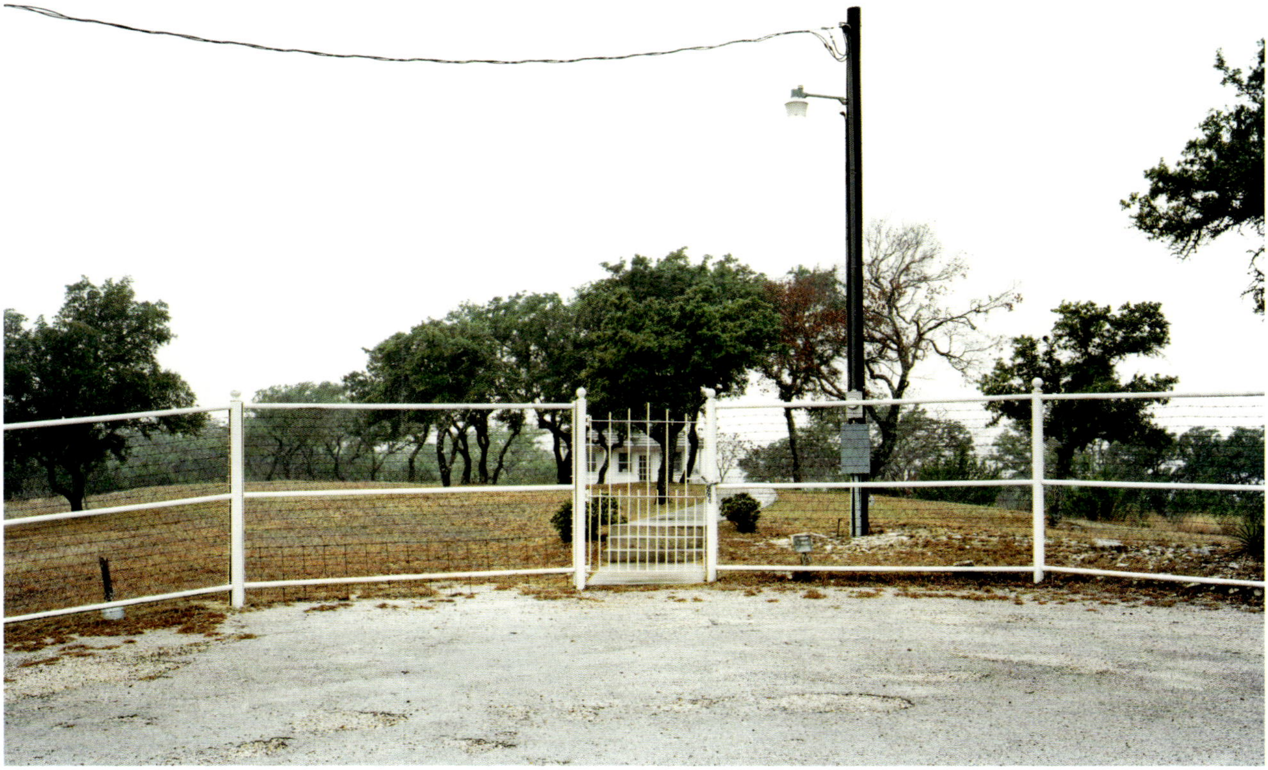

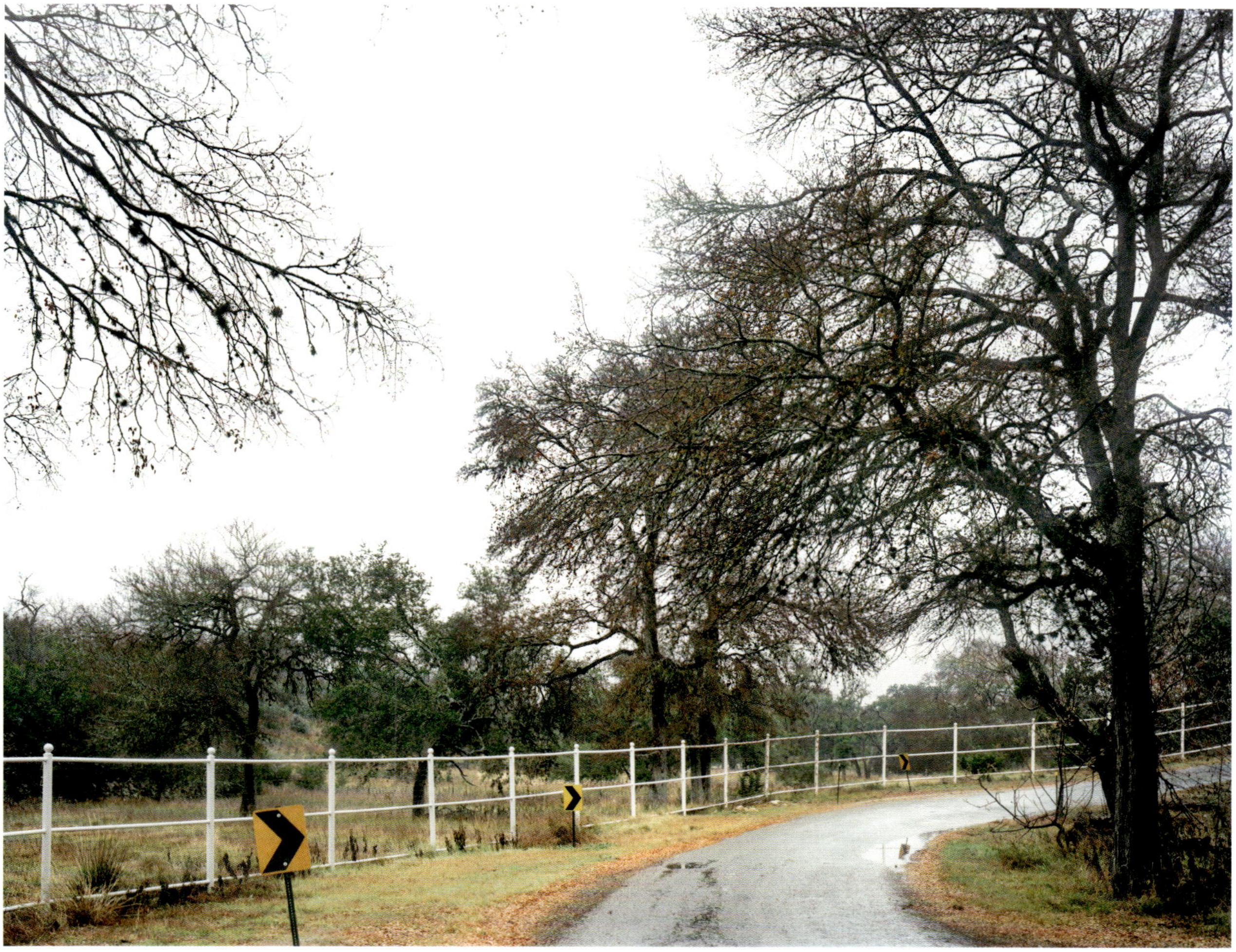

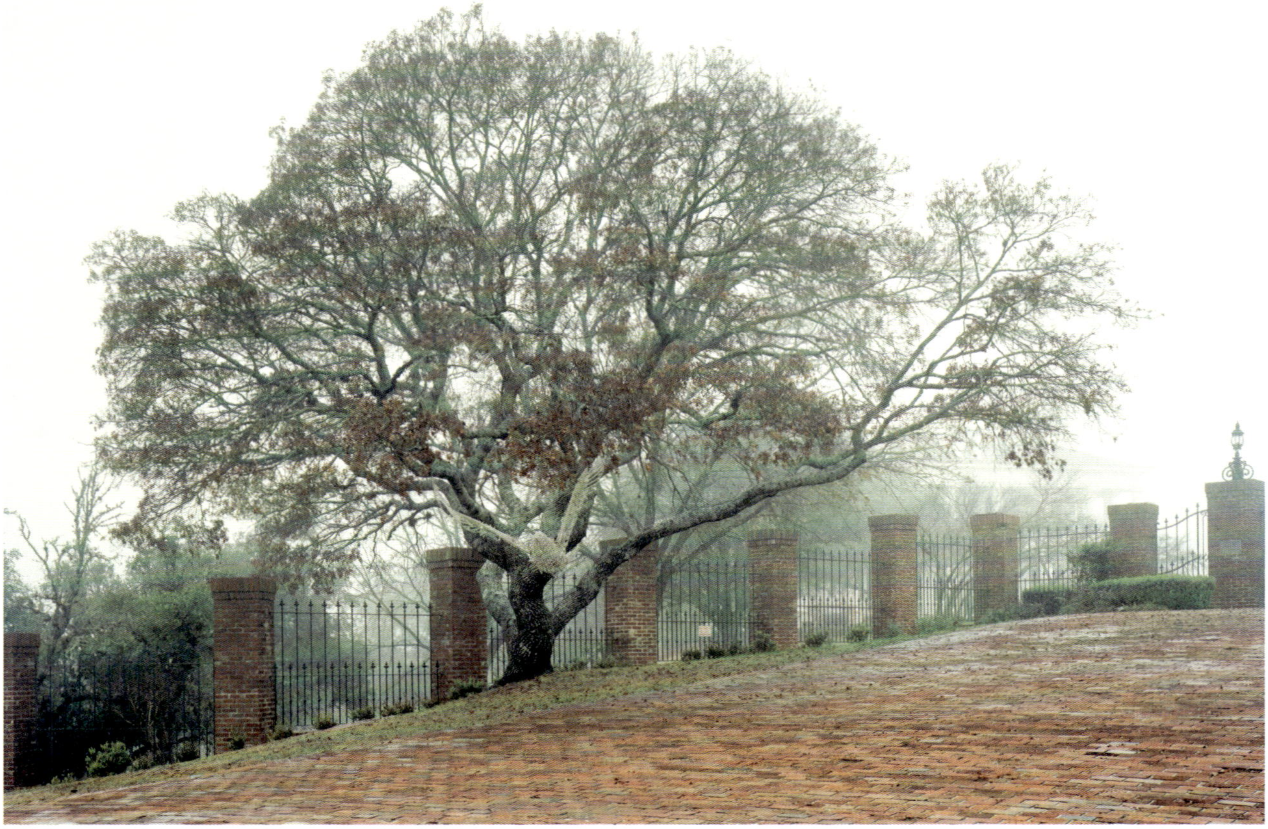

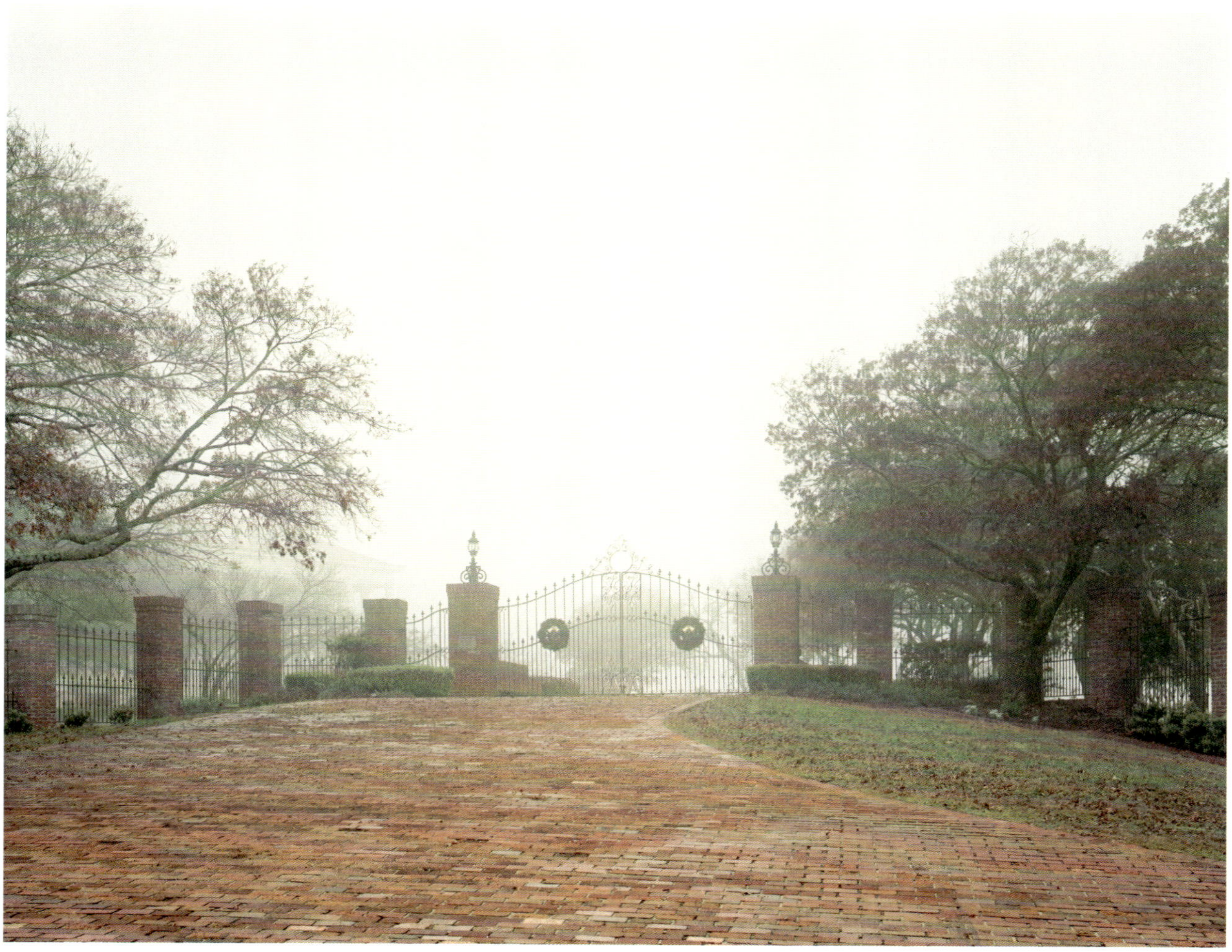

TED'S
ARCHIVE

PHOENIX, ARIZONA

Ted Kraver graduated from MIT in the top ten per cent in his class, with degrees in aeronautical and astronomical engineering. He went on to work as an engineer for Garrett-AiResearch and contributed towards the NASA Saturn V rocket project.

In 1966, Ted, along with fellow MIT graduate Frank Reichenbacher and local entrepreneur and wig maker, Ed Hope, established an independent cryonics company called Cryo-Care.

In 1969, after realising that the company was not financially viable, the partners liquidated the company.

Along with his wife Bobbie, Ted continues his affiliation to the Alcor Life Extension Foundation and will be cryo-preserved upon his death.

Ted kept his photographs of Cryo-Care's short history for posterity.

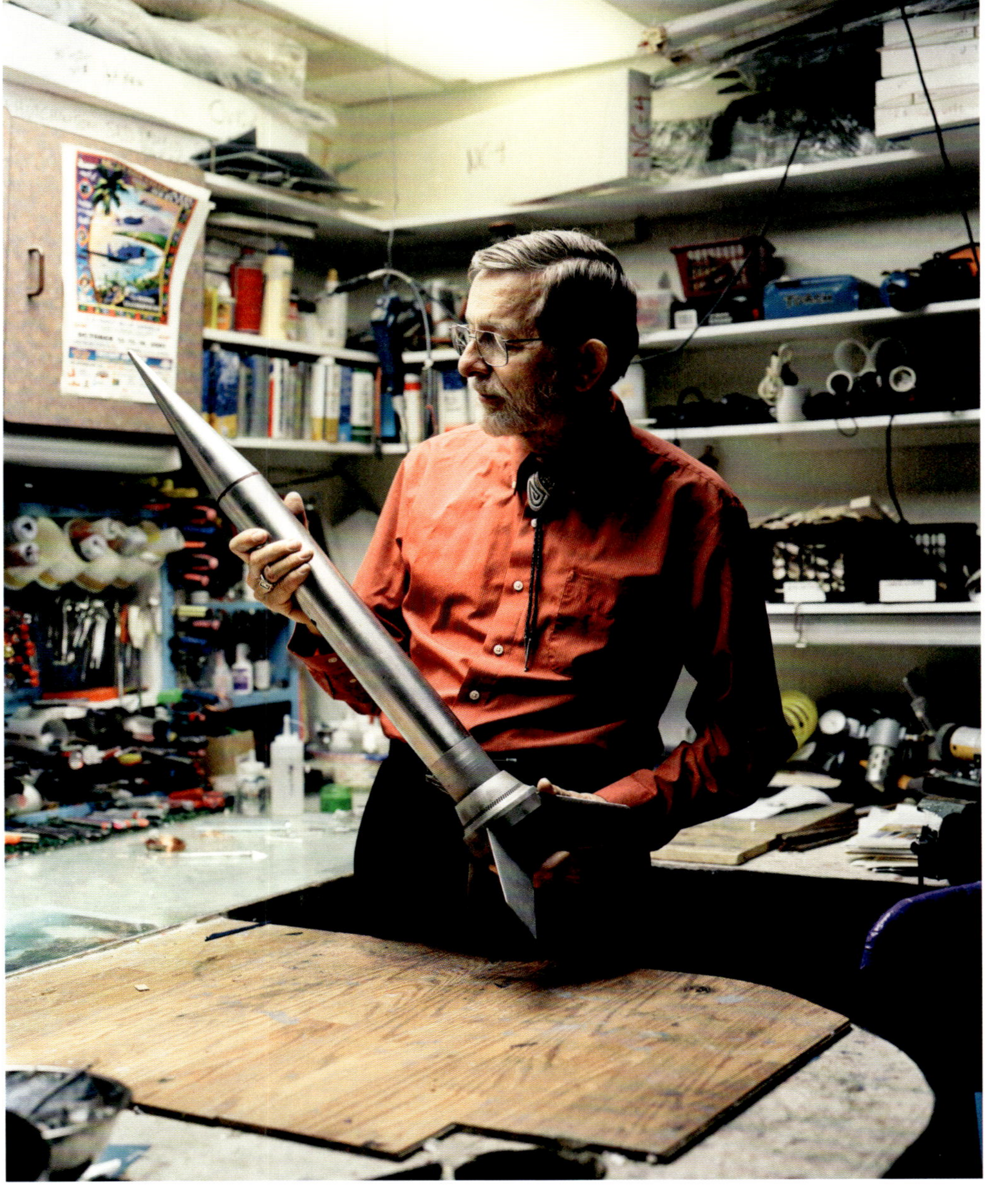

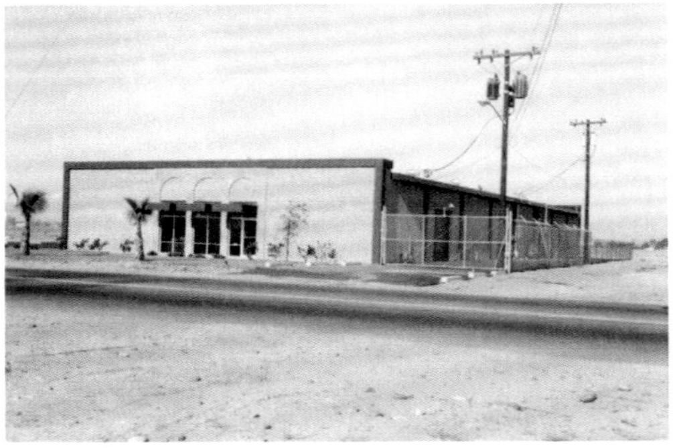

1

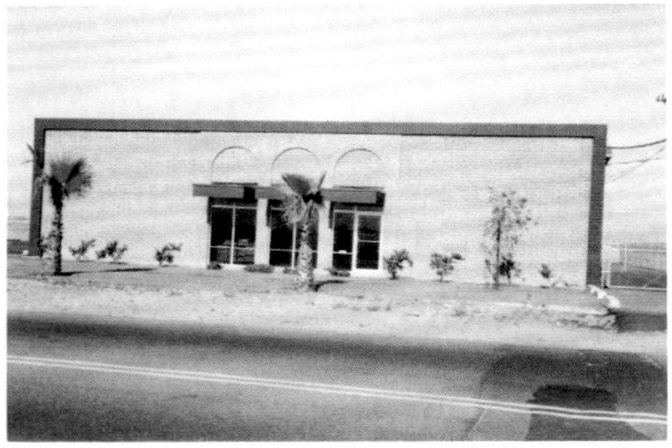

2

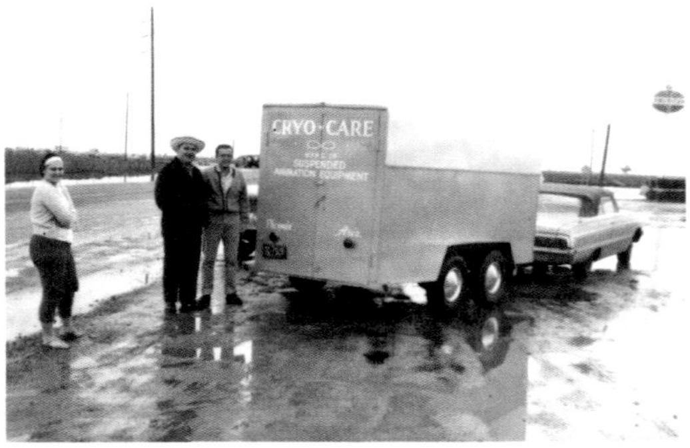
CRYO-CARE
SUSPENDED
ANIMATION EQUIPMENT

3

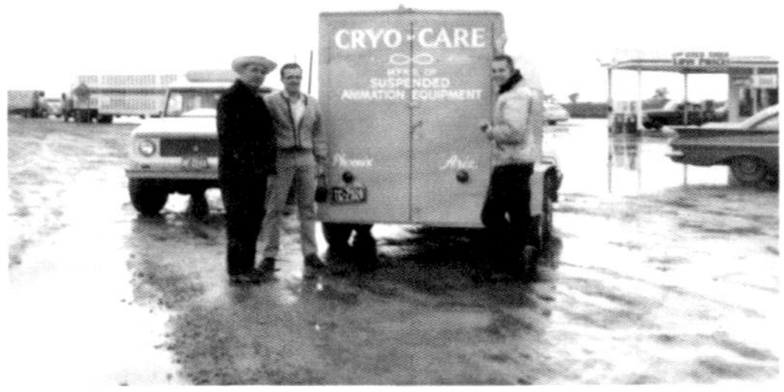
CRYO-CARE
SUSPENDED
ANIMATION EQUIPMENT

4

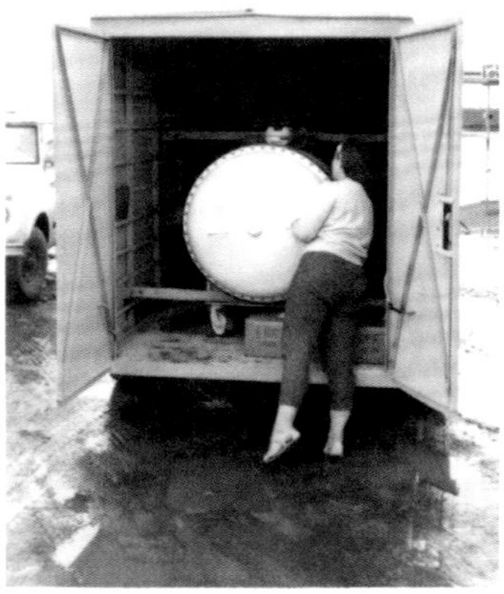

5

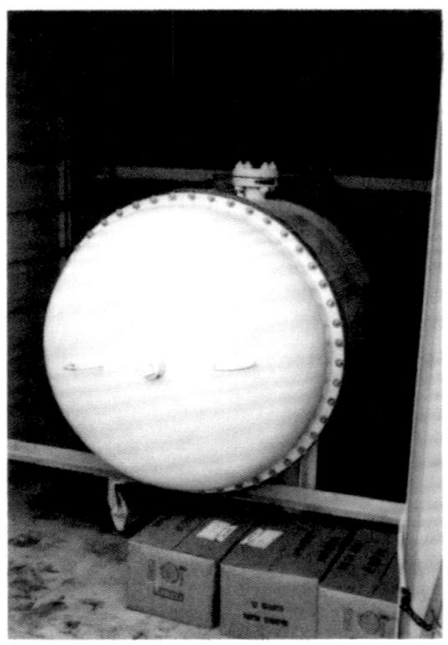

6

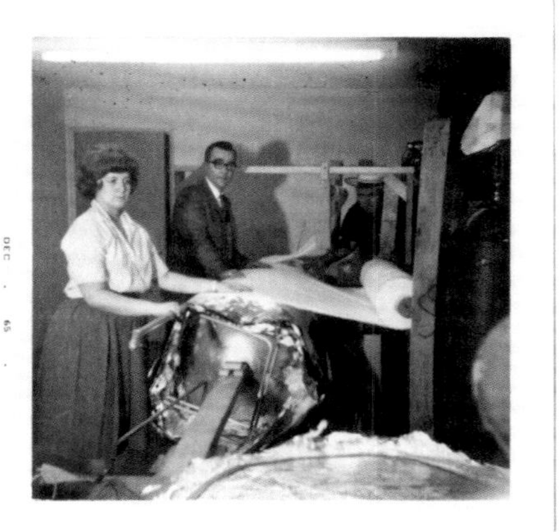

7

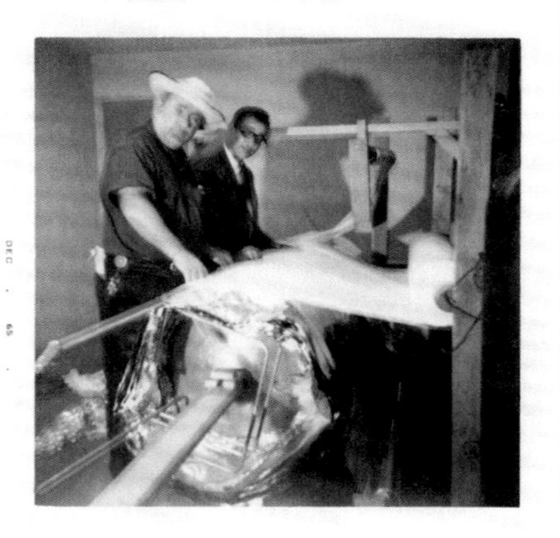

8

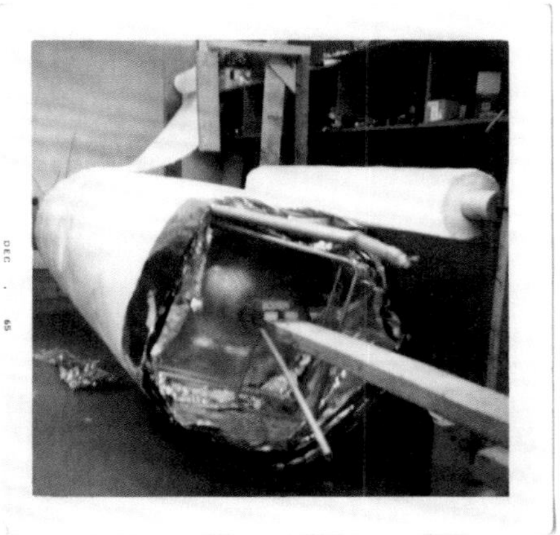

9

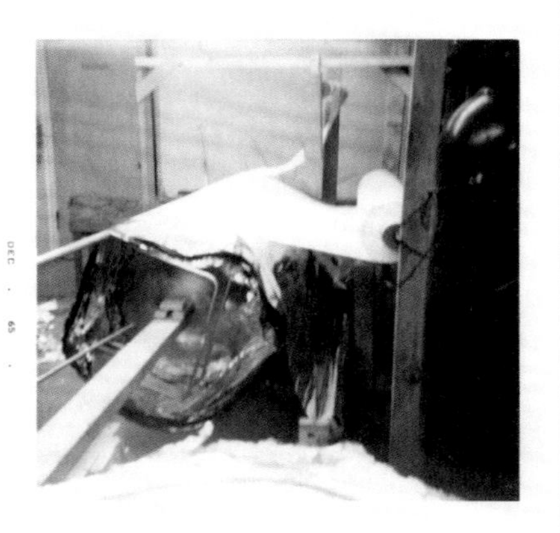

10

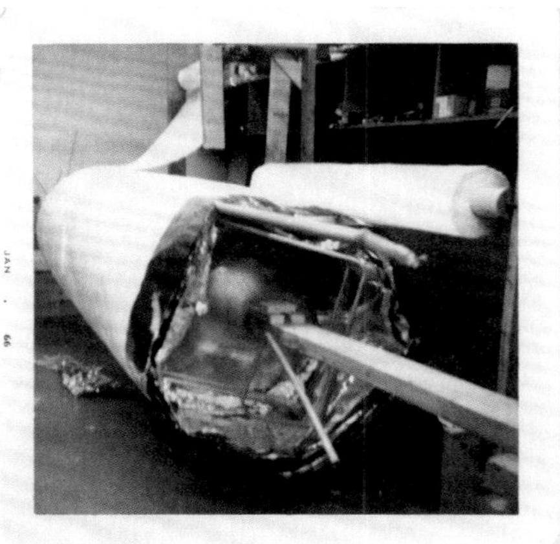

11

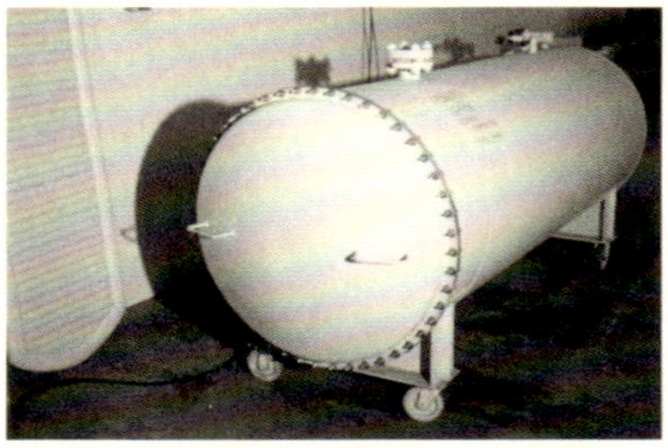

12

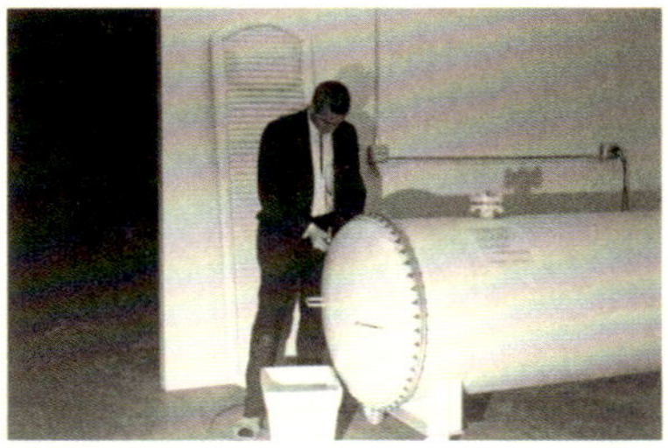

13

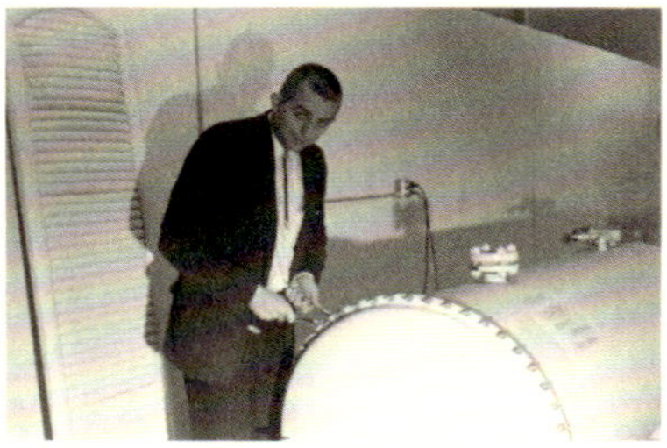

14

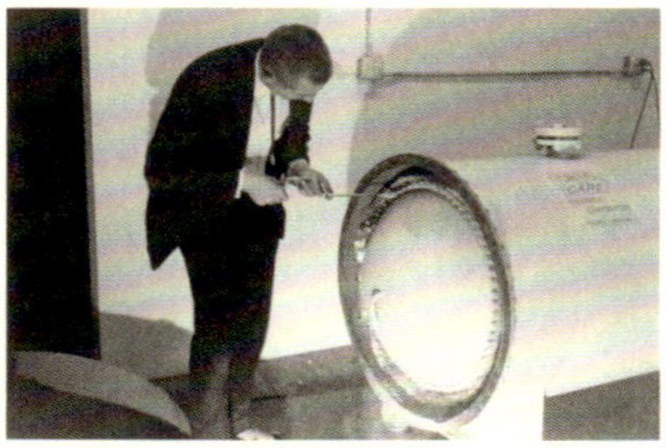

15

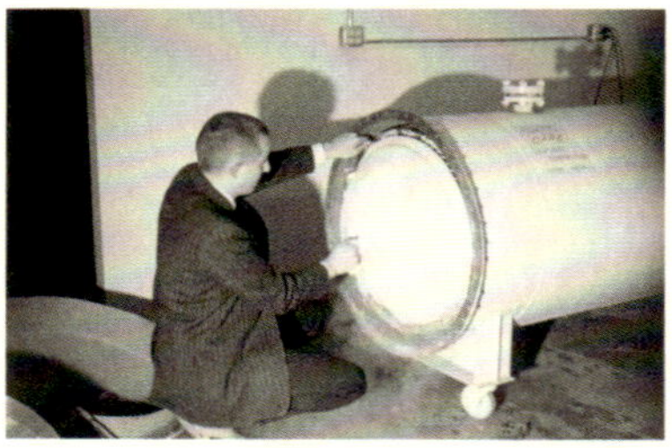

16

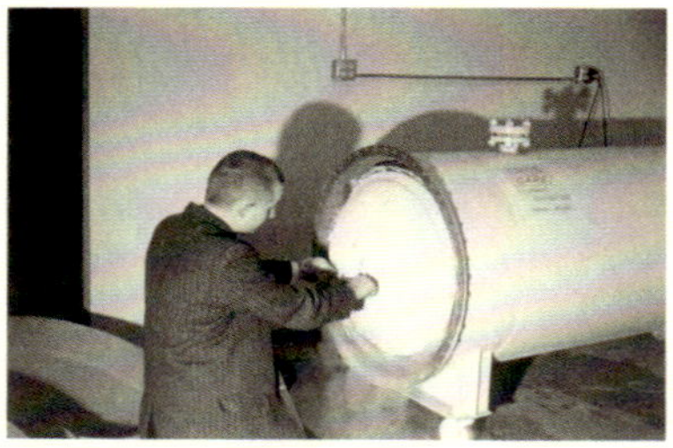

17

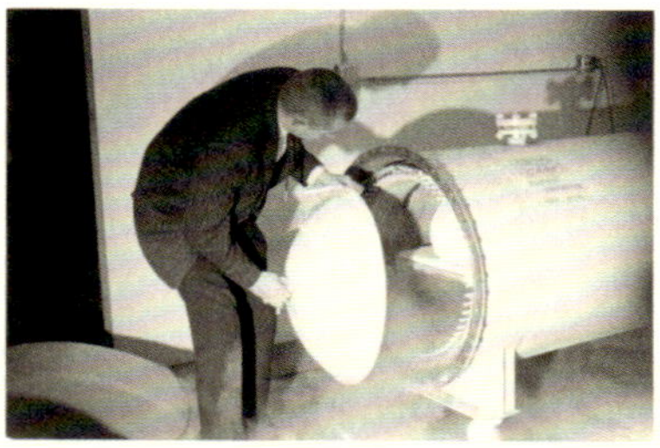

18

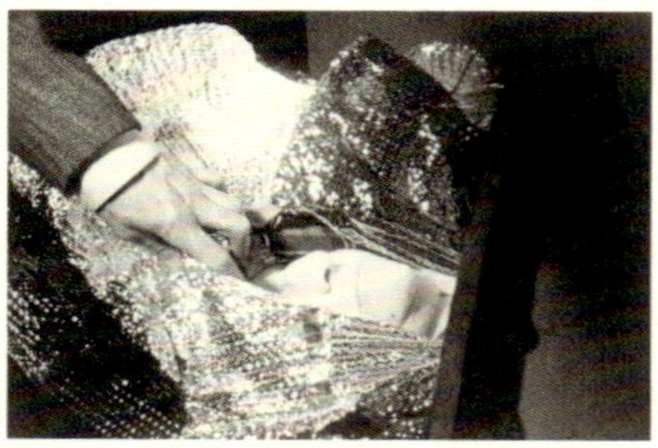

19

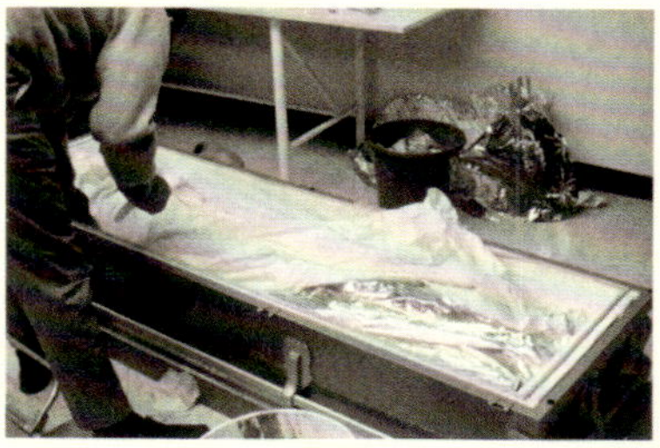
20

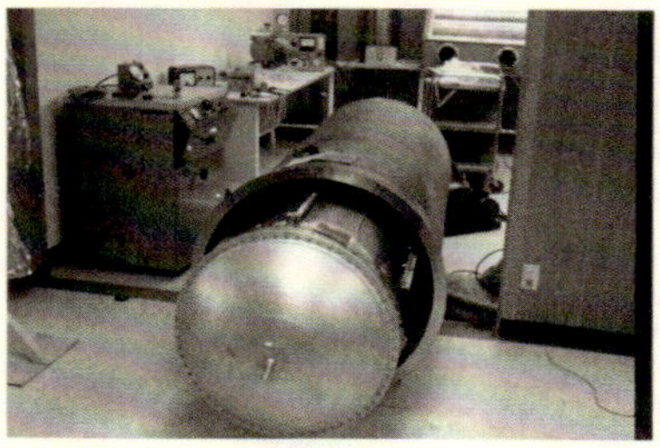
21

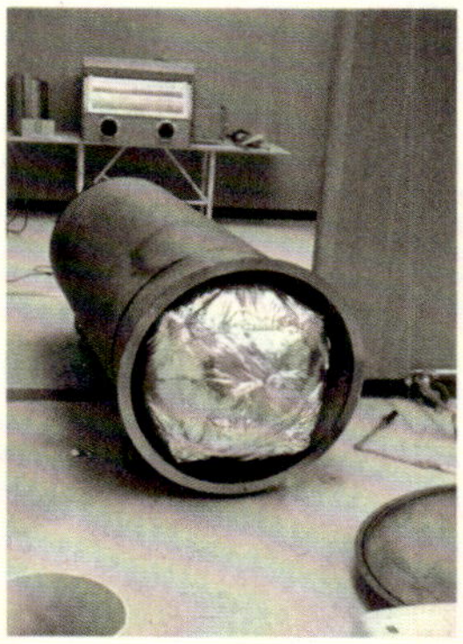
22

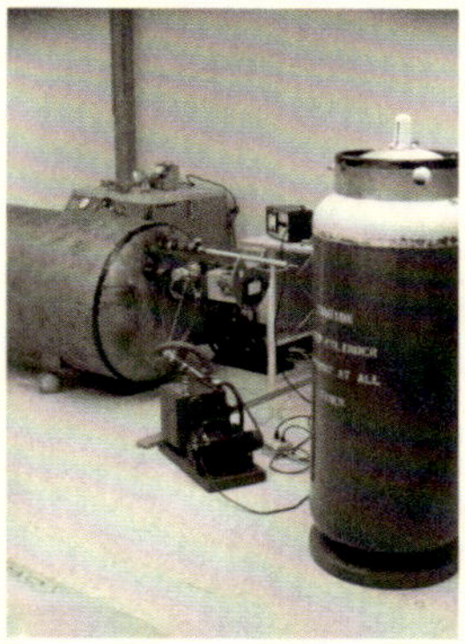
23

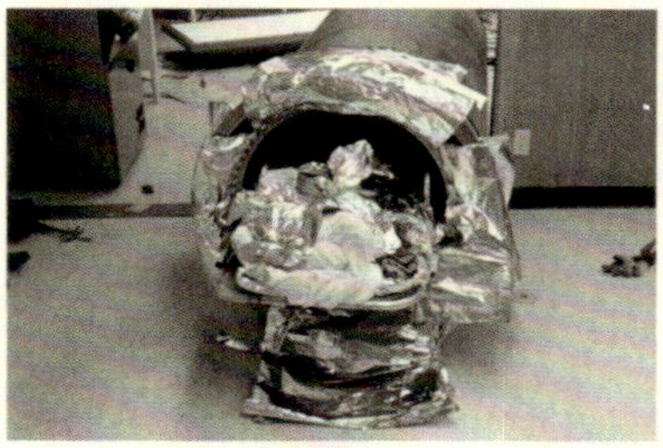
24

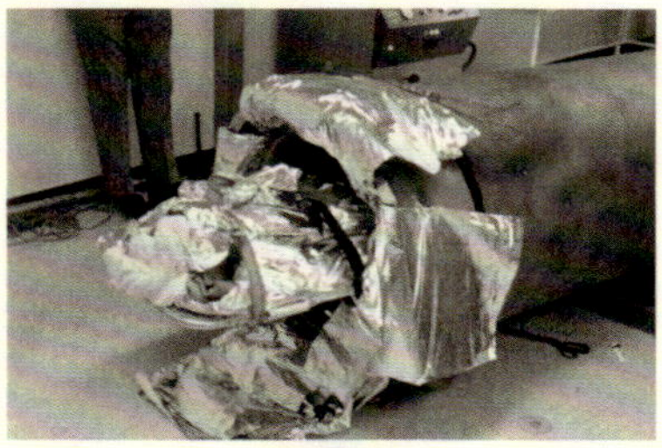
25

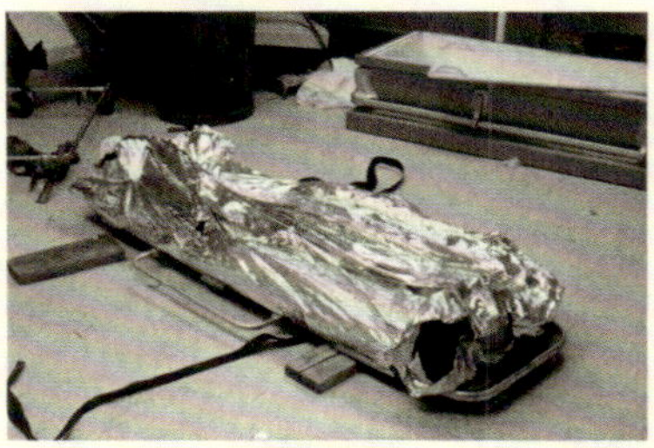
26

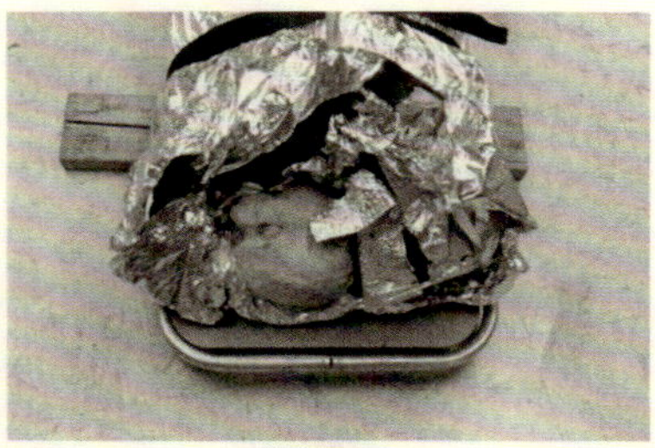
27

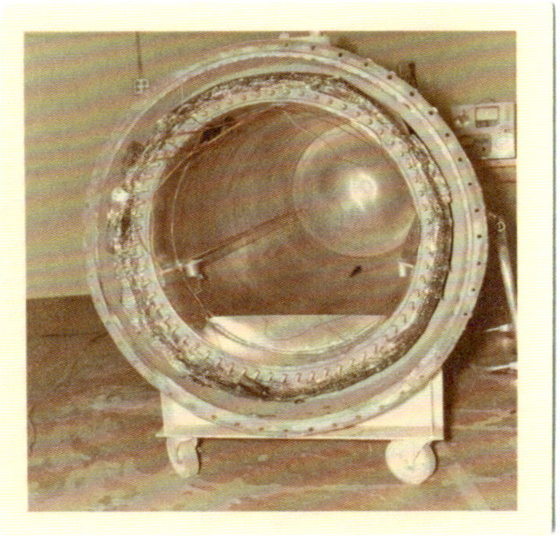

28

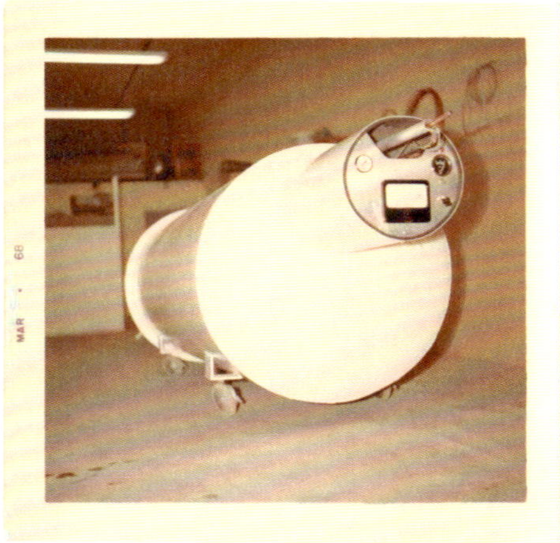

29

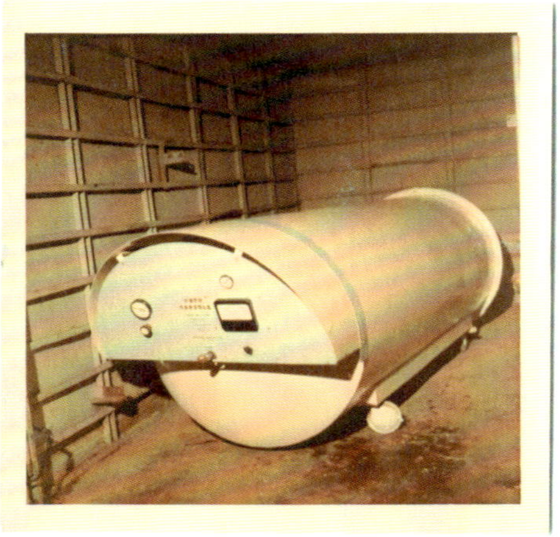

30

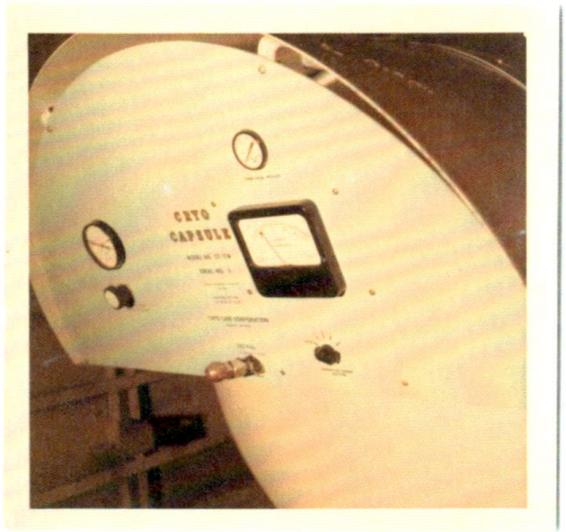

31

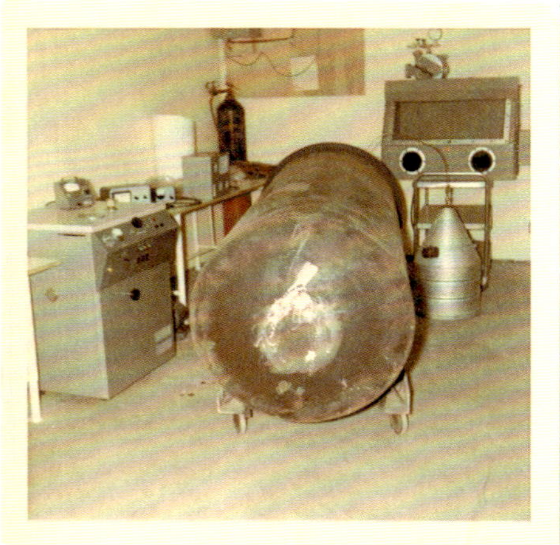

32

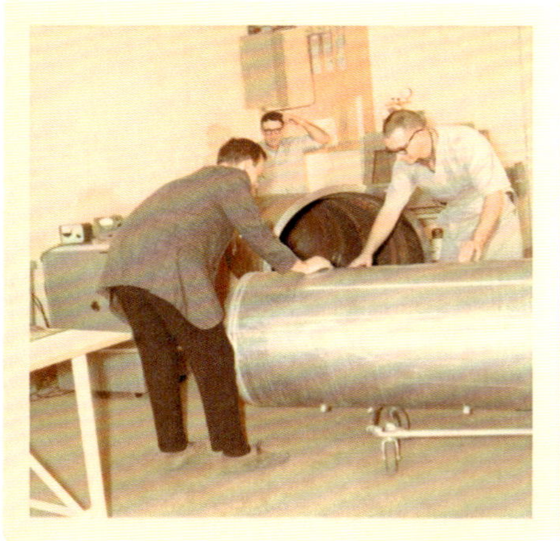

33

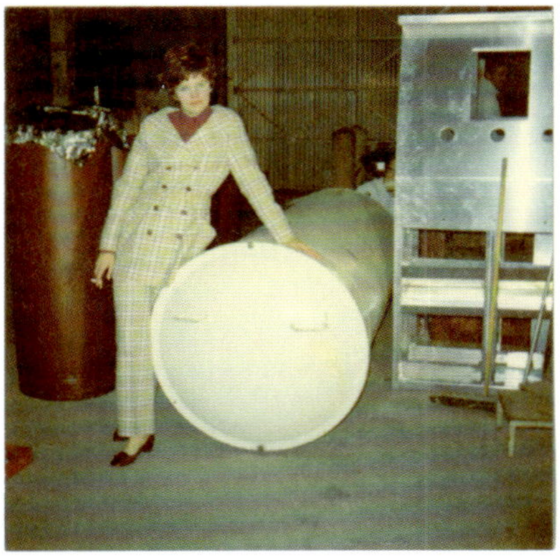
34

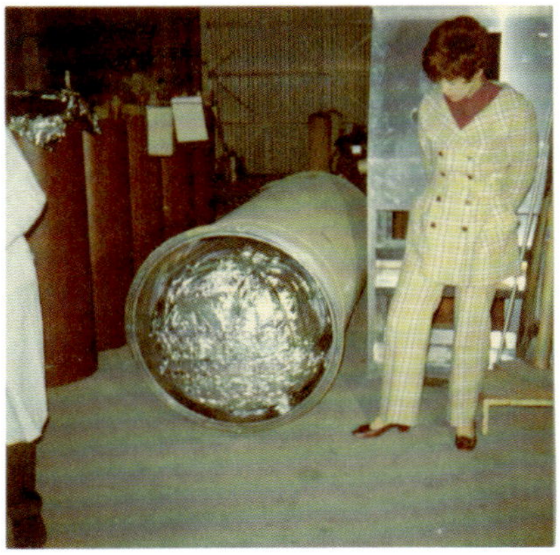
35

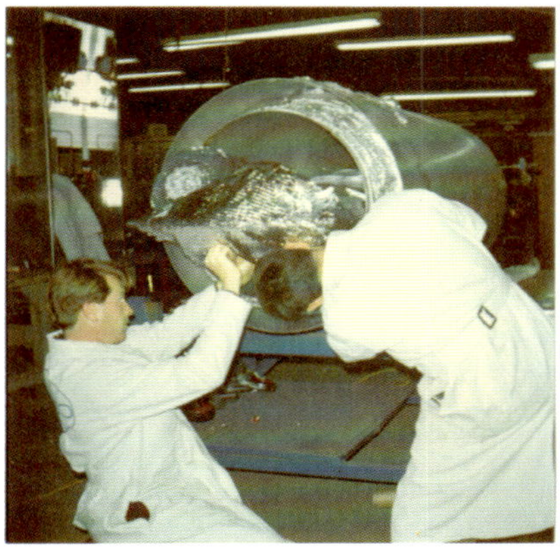
36

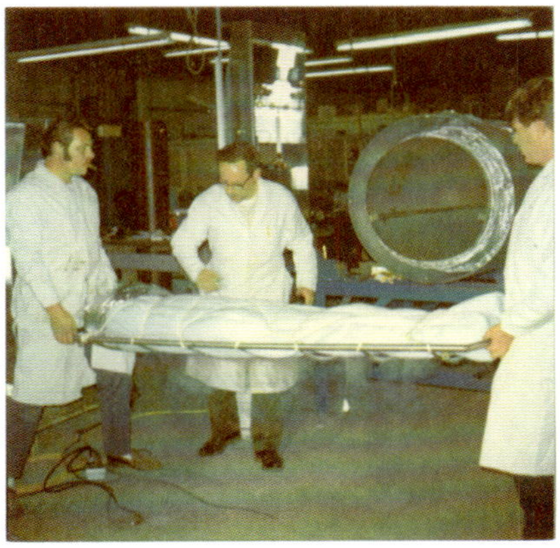
37

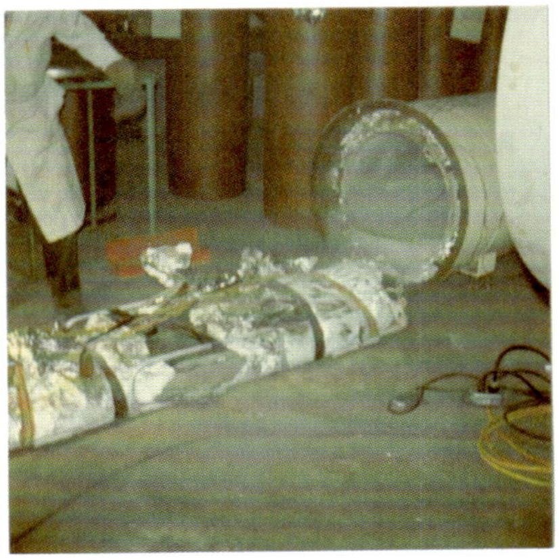
38

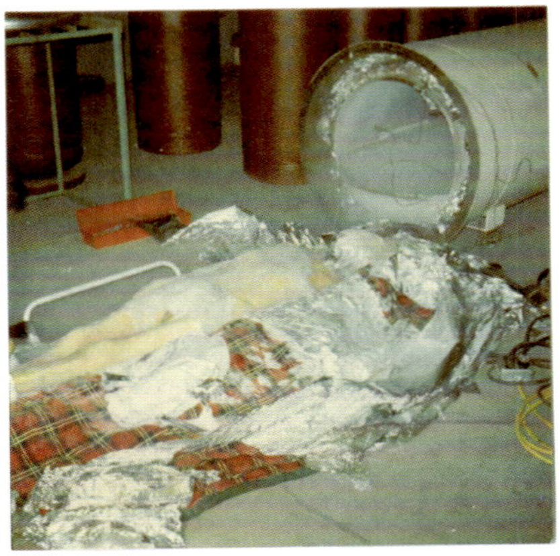
39

NOTES

In 1962 Robert C.W. Ettinger published *The Prospect of Immortality*, the book largely credited with launching the cryonics movement – the process of freezing a human body after death in the hope that scientific advances might one day restore life.

In the foreword Ettinger wrote: 'This book is intended to convince first a substantial number of scientists and intelligent laymen, and then the nation, and eventually the world, that the prospect of immortality is not idle speculation. Rather, it represents an issue, which is 'vital' in a wide and literal sense, one which should soon have sledgehammer impact on every facet of personal and national life, and which urgently requires action by all of us as individuals'.

This publication is the result of Murray Ballard's ten-year photographic study of the cryonics industry, its technology, the characters involved and the facts and myths surrounding this controversial subject.

Robert C.W. Ettinger's Micrographs; **Front endpapers and p1-4**
These micrographs were found amongst Ettinger's possessions at the Cryonics Institute. It is thought they date from the 1970s when he was conducting research into vitrification and how it could minimise freezing damage in body tissue. Unfortunately his notes could not be found so it is difficult to determine the results.

Vitrification is cryopreservation, which is done with cryoprotectant (antifreeze) mixtures. The aim is to eliminate freezing – ice crystal formation. In vitrification, tissues of the cryonics patient harden like glass rather than crystallise. It is currently not possible to vitrify most organs, but the brain can be vitrified.

The Martinot Story
Nueil-sur-Layon, Pays de la Loire

L'Inconnue de la Seine, Raymond Martinot's study. April 2007; **p9**
L'Inconnue de la Seine (The Unknown Woman of the Seine) was an unidentified young woman whose anonymous death mask became a popular fixture on the walls of artists' homes after 1900. Her face inspired numerous literary and artistic works.

According to the story, a young woman's body was pulled out of the Seine River at Quai du Louvre in Paris around the late 1880s. Since the body showed no signs of violence, suicide was suspected.

A Paris morgue pathologist was so taken by her beauty that he made a wax cast death mask. In the following years, numerous copies were produced. The copies quickly became a fashionable morbid fixture in Parisian Bohemian society. Her enigmatic smile has been compared to that of the Mona Lisa, inviting numerous speculations as to what clues the eerily happy expression in her face could offer about her life, her death and her place in society.

Le Chalet de Preuil (view from the road). March 2007; **p11**

Bust of Raymond Martinot. April 2007; **p13**

Madame Monique Leroy's bedroom. April 2007; **p15**
Whilst the experiment was going on, Raymond, and then Rémy, kept Monique's bedroom exactly how it was for her to return to. When the experiment came to an end, Rémy started using his mother's bedroom as the château fell into disrepair.

Freezer. April 2007; **p16-17**
The decision to put the freezer in the cellar underneath the château was made after bargaining with the local authorities. It was put in place by six or seven men and only remains there, now the experiment has come to an end, because it is very difficult to move.

Rémy Martinot-Leroy. April 2007; **p19**
Raymond Martinot and Monique Leroy's only child, Rémy, lives just outside of Paris and comes down to stay at the château sporadically throughout the year. Raymond wanted Rémy to live permanently at the château whilst he was alive in case he died unexpectedly. He also asked him to live there after he died to keep an eye on the experiment, in case something went wrong, but Rémy refused. He admitted the experiment was a considerable burden and he did not want to be the 'bishop of this church'.

Cryoprotectant. April 2007; **p20**
A bottle of cryoprotectant used in the preservation of Raymond Martinot.

Cryoprotectants are anti-freeze substances that prevent or reduce ice crystal formation.

Torn Tag. April 2007; **p21**

Ashes in the garden, Le Chalet de Preuil. March 2007; **p23**

Alcor Life Extension Foundation
Scottsdale, Arizona

Alcor Life Extension Foundation (front view). December 2012; **p27**
In 1972, Alcor was first incorporated as the Alcor Society for Solid State Hypothermia by Fred and Linda Chamberlain. On 16 July 1976, Alcor performed its first human cryo-preservation. In 1977 the name was changed to Alcor Life Extension Foundation.

Meeting for prospective members. October 2006; **p29**
Readiness Coordinator Michelle Fry explains the cryo-suspension procedure to a group of prospective members.

It costs $200,000 to have your whole body cryo-preserved and stored indefinitely at Alcor or $28,000 for neuro-patients. A neuro-suspension is a cryonics patient for whom only the head has been cryo-preserved.

Aaron Drake, Medical Response Director, prepares stabilisation medications to be used during the initial stages of a cryopreservation. August 2009; **p30**
Once a cryonics patient is pronounced legally dead, the response team place

them in a bath of ice to start cooling the body. An automated heart-lung machine is used to keep the cells of the patient's organs and tissues alive and restore blood flow needed to administer anticoagulants and medications. It is very important these processes are carried out correctly, otherwise perfusion (blood washout) and vitrification (insertion of cryo-protectant) may not be possible.

Operating room. October 2006; **p31**
Once they have died, cryonics patients are brought to the operating room to undergo the procedure of vitrification. Vitrification is cryopreservation done with cryoprotectant mixtures, which are so effective that freezing can be eliminated entirely. In vitrification, tissues of the cryonics patient harden like glass rather than crystallise. At this stage the body is cooled, using liquid nitrogen, to prevent it deteriorating.

Patient storage demonstration. August 2009; **p33**
Bodies are stored in sleeping bags and strapped into stainless steel cases to protect them during long-term storage. As the brain is the most important part of the body, patients are stored upside down with their head inside a separate container.

Large cooldown lid. April 2007; **p34**
This device is used in the final stage of a cryo-suspension to cover the dewar in which a whole body patient is cooled to -195.76°C (liquid nitrogen temperature). It is important cooling takes place gradually in order to prevent the patient from cracking.

Hugh Hixon, facilities engineer and research fellow. April 2007; **p35**
Hugh has worked at the cryonics facility as an engineer since 1982. He is in charge of technology application and development. He designed and constructed the majority of equipment used at the facility, including the 'patient care bulk-fill system' and the 'large cooldown lid'.

Patient Care Bay (Bigfoot dewar being filled with liquid nitrogen). October 2006; **p36-37**
It's often impossible to predict when a cryonic suspension will take place, so the facility needs to be ready to receive patients at all times. At Alcor, patients are stored permanently in dewars, which work in the same way as a thermos flask and are not dependent upon electricity.

Aido's toy and membership bracelet. April 2007; **p38**
Dr Mike Perry's cat – and Alcor's unofficial mascot. Aido, died in July 2012 aged 20 and was subsequently cryo-preserved. He is the only perfused cat in storage.

Dr Mike Perry, patient caretaker. April 2007; **p39**
Mike lives on site at the facility; his main job is to monitor the patients in storage. He has been a member of Alcor since 1984.

> *All my life I have been interested in some way of getting around the problem of death. At first, I was taught about supernatural powers that were able to resurrect the dead, a feat no human science supposedly would ever accomplish or explain. But my belief in the supernatural faded out when I was a teenager, so I became interested in alternatives. Cryonics is the best I have found so far.*

Liquid nitrogen delivery. August 2009; **p40**
Liquid nitrogen (LN_2) is normally delivered once a week. Due to a small amount of boil-off, the patient dewars are also replenished with LN_2 once a week. In an emergency, it is estimated the dewars could sustain cooling for 90 days before the LN_2 would be exhausted and catastrophic warming would occur.

Cryonics UK
Peacehaven/London/Sheffield

Frank, prospective patient. Standby team training. Peacehaven, East Sussex. May 2007; **p45**
Frank is a software engineer from Cambridge who signed up as a prospective patient in 2005. It is estimated that 80 per cent of those interested in cryonics are employed in this field.

Bob (practice body), standby team training. Sheffield, South Yorkshire. July 2010; **p46-47**
As there are no cryonics storage facilities in the UK, the initial stages of a cryopreservation are normally carried out at a hospital, or occasionally at a patient's home, before being packed in an insulated shipping container and flown to the USA.

Medications, standby team training. Peacehaven, East Sussex. December 2006; **p49**

Standby team training, Sheffield, South Yorkshire. July 2010; **p51**
Volunteers meet over a weekend two or three times a year to practice the initial stages of a cryopreservation. These include:

Initial cooldown using water ice while providing CPS (cardio-pulmonary support).

Blood washout, replacing blood with organ-preservation solution.

Perfusion with cryoprotectant (human anti-freeze solution).

Packing the patient into a shipping container with dry ice.

Maria Camacho, prospective patient, at her home in Brixton, London. April 2011; **p53**
Maria has been interested in cryonics ever since she was a young girl growing up in Colombia during the 1960s. In 1968, her father, a doctor who specialised in radiology, self-published a science-fiction novel called *Siete Minutos* (*Seven Minutes*). In the story, a man named Homer becomes a millionaire, then the sun explodes and he's left as the only survivor on Earth, until he finds and revives a cryonically-preserved man. Maria moved to the UK in 1975 to learn English and study creative writing with the aim of translating her father's novel. In 1995, she signed up as a prospective patient with the Cryonics Institute. Maria's father did not.

Maria's children's height markings. Brixton, London. March 2009; **p55**
Maria has three children. Her son Michael signed up when he was 18 years old and daughter Romina signed up when she was 16. Her other daughter, Rosanna, decided against it.

Portable perfusion kit, standby team training. Peacehaven, East Sussex. December 2006; **p56-57**

DNA archive. Home of David and Ellen Styles, Macclesfield, County. February 2009; **p59**

> *It's an absolute last-ditch effort – the last safety net, the last resort – if cryonics were to be impossible, such as if one was blown to bits in an air crash or such like. At the very least, one could be cloned, the clone can have memories put into it, if you have journaled effectively, and if enough is known about your life, people cannot tell the difference between a real memory and an implanted one; that's actually a surprisingly easy thing to do. Whether this constitutes a kind of immortality or whether it's simply creating a copy who might believe they are you is a question for philosophers. In this case, it's something that's very cheap, quick and easy to do. There's no real downside to doing it, and maybe it's relevant, maybe it's not, but we like to be able to say, at least we tried.*
> David Styles

KrioRus Part 1
Moscow/St Petersburg

Danila Medvedev, KrioRus founder. Outside the KrioRus office, Moscow. February 2010; **p63**
Danila translated Robert C.W. Ettinger's *The Prospect of Immortality* into Russian. Medvedev is a prominent Russian cryonicist and member of the Russian Transhumanist Movement.

> *Cryonics is very useful to the transhumanist movement in Russia because, even though cryonics itself is not very credible still in the world – not mainstream – transhumanism has even bigger problems. It doesn't have anything concrete to offer that is impressive, that is already physically present and available, something you can touch. It's all about the ideas of the future and with cryonics we can at least have one concrete example of what we, as transhumanists, can do. So when people hear our stories about singularity, immortality, et cetera, they can't dismiss it as just talk, because at least we can show that cryonics is here and cryonics is available in Russia.*

Transhumanist meeting. Kriorus office, Moscow, Russia. February 2010; **p64-65**
Transhumanism is an international movement supporting the use of science and technology to fundamentally transform the human condition by developing and making widely available technologies to eliminate ageing and to greatly enhance human intellectual, physical and psychological capacities.

Valerija Pride, General Director. KrioRus facility, Alabushevo, Moscow. April 2009; **p67**
Since 1990, people have wanted to start a cryonics organisation in Russia, but nothing was done for 14 years. KrioRus was launched on 3 May 2006.

Photographs of Lidia Fedorenko, the first cryonics patient in Russia. St Petersburg, April 2009; **p69**

Ex-military surgical equipment. Eugene's apartment, St Petersburg. April 2009; **p71**

Medications. KrioRus office, Moscow. April 2009; **p73**

Dewar. KrioRus facility, Alabushevo, Moscow, Russia. September 2010; **p75**
Stores the head of the first cryonics patient in Russia, Lidia Fedorenko, alongside numerous DNA samples.

Lidia Fedorenko's grave. St Petersburg, April 2009; **p76-77**
In 2005 Galina's mother, Lidia, was the first cryonics patient in Russia. She is a neuro-patient so only her head is cryonically preserved; her body was buried in this cemetery on the outskirts of St Petersburg. The theory behind neuro-patients is that only the information contained in the brain is of any importance, and that a body to contain the revived brain could be cloned or regenerated at some point in the future.

Cryonics Institute
Clinton Township, Michigan

John Bull, a prospective patient, records a video instructing people what to do upon his death. Terasem Movement Inc. Melbourne Beach, Florida. April 2010; **p81**
One of the major obstacles to successful cryopreservation has been disputes between relatives. Any such dispute usually blocks the ability of the cryonics organisation to use vitrification, and consequently dramatically reduces the probability of ever successfully bringing a person out of a cryo-preserved state. Therefore, some members decide to record video testimonies to make their wishes clear. John Bull was cryo-preserved in January 2014.

Cooling chamber. March 2007; **p83**

Cryostats. April 2010; **p84-85**
A cryostat is defined as an apparatus used to maintain constant low temperature. It is also the word used by the Cryonics Institute to describe the containers used to store cryonics patients. Six patients are stored in each one.

Flower box. April 2010; **p87**
Patients' families and friends leave flowers in the numbered slots, corresponding with where their loved ones are held.

> *For Rosie, we will miss you. Hope the flowers remind you of Spring. We hope to see you again, ❤ Mike*

Portraits of patients currently held in cryostasis. December 2012; **p88-90**
The Cryonics Institute only accepts whole-body patients. It costs $28,000 as a one-time payment, although most members pay through life insurance. In December 2012 the Cryonics Institute had 113 patients in cryostasis.

Life Extension box. December 2012; **p91**

Liquid nitrogen level measure. March 2007; **p93**

Catstat. March 2007; **p95**

Bed-alarm system. April 2010; **p97**
Ben Best explained that the most important thing for a successful cryonics suspension is having your body preserved as quickly as possible after you die. So one of the biggest hurdles is to avoid dying at home in your sleep, because it could mean your body isn't discovered for some time. Ben's solution was to develop a bed alarm system.

A Place of Last Stay
Roseville, Michigan

George Vogler. Cryonics Institute, Clinton Township, Michigan, USA. April 2010; **p101**
George was at the Cryonics Institute manning the phones. He had been involved with the cryonics movement since the latter part of the 1960s, but it was only recently, after retiring, that he was able to be more actively involved.

George in his backyard. April 2010; **p102**

> *Right now, the house has two fully functional bedrooms on the main floor and one on the top floor. It was originally set up to have two basement bedrooms. I'm presently not using it that way. It also has a bath that only has the shower section and isn't otherwise completed. I'll try and get that set up so the basement has functional bedrooms. The section on the rear of the house, which is under a roof, I'd like to make that into a sort of a library – I'm not sure of a good word for it, a room where we can have entertainment or people who want to have meetings or things of this sort – as well as making it the entrance way to the one bedroom that will have a ramp and whatnot for people in wheelchairs or rolling beds. After I bought the house, the city decided to condemn it. This is not a problem with the house so much; it's an economic problem. The cities here are trying to put the Union workers to work. Unfortunately, I haven't been able to support many of the Union labourers; I'm doing the work myself.*
> George Vogler

George's bed. April 2010; **p104-105**
Including bed alarm system.

Videotapes, Fred's bedroom. December 2012; **p107**

Fred, George's house. December 2012; **p108-109**
Fred had signed up to the Cryonics Institute eight years previously. He had come from Dayton, Ohio to live with George so he was closer to the facility.

Let me say this: my name is Fred Coates. I'm 80 years old. I'm older than television, I'm older than frozen food... Years ago, I don't know what year, but the person by the name of Mary Shelley wrote a story about a doctor named Frankenstein and he created a monster. Now what Dr Frankenstein would say to people is...'bring me your bodies, bring me your bodies'. From which he constructed a monster, or what was considered a monster… They made movies of it... Now people used to think that that was scary, but today, when you go get a driver's license, what do they ask you? They ask you... Bring me your organs, bring me your bodies, we want your organs. It's not scary any more; it's technology now… The Wright brothers in 1903 were the first heavier-than-air flight and, only 66 years later, a guy named Neil Armstrong walked on the moon, the first human on another planet. Now, everything that you can think of that man can do, he can achieve one way or another… I'm 80 years old, I've abused my heart by smoking for 50 years and drinking tequila – a litre every two days. But the heart is a wonderful piece of machinery and it's still working after 80 years and all the abuse, it's still working… It's a wonderful piece of machinery. But, only 30 years ago, if a man had a heart attack he died on the street and there was no such thing as artificial respiration, CPR… Then they also developed heart transplants… Now they have the heart thing completely under control. And they're also starting on cancer… In the near future, there's going to be such a thing called nanobots, which are extremely small robots in your blood stream, extracting the bad things that harm you. Today, you take medicine to kill those things that can harm you, but the chemicals do as much harm, or more, than what they cure. I see ads on TV… take so and so for some simple thing, but it will have 50 reasons why not to take it… So with nanobots the future is going to be a great place to live… Who knows what the future will bring. Now the next few years, maybe 10, 15, 20 years, will be the only period of time when people are frozen except when they travel in outer space. If you travel in outer space you might have to be frozen for a few thousand years unless they can find a quicker way to get to another part of the universe. But some day it may be against the law not to have somebody frozen or cured. In 20 years they'll be able to cure everything.
Fred Coates

Fred's suitcase. December 2012; **p110-111**
Inside the suitcase were all of Fred's personal photographs. His life recorded in snaps, from his first drum kit to a framed photograph of his wife who died a few years ago. She was not cryonically preserved.

Grandpa Bredo
Nederland, Colorado/Oslo

View from the Rocky Mountain Cryonics Facility, overlooking Nederland and the Barker Reservoir. March 2007; **p115**

Rocky Mountain Cryonics Facility. March 2007; **p117**

> *You will be aware that there might be one world war every century and there might be, if not nuclear war, a war that's worse. So, I drew up designs for how one could build a nuclear war-proof, earthquake proof, cryonics laboratory and storage facility.*
> Trygve Bauge

Checking the dry-ice levels. December 2012; **p119**

> *I go down to Boulder once a month and pick up about 1800 pounds of dry ice, bring it up and fill up Grandpa Bredo. And then I come up every week thereafter and check on the ice levels. And make sure that the ice is not melting at a higher rate or that the case is not broken.*
> Jane, Grandpa Bredo's caretaker

Looking into Grandpa Bredo's storage box. December 2012; **p121**
Grandpa Bredo died on 6 November 1989 in Oslo.

> *My mother called. He had died during his nap. They found him 10 minutes or 15 minutes after he died. I was home in Boulder, Colorado; I got it and I immediately said to my mother, we have to freeze him. My sister was vehemently opposed to it, but we ignored her completely.*
> Trygve Bauge

Grandpa Bredo's body arrived in the USA on 1 February.

Residential area, Rocky Mountain Cryonics Facility. December 2012; **p122-123**

Aud. Oslo, Norway. April 2015; **p125**

Trygve ice-bathing. Heggelielva, Sørkedalen, Norway. May 2015; **p126-127**

> *I started a New Year's Day tradition. Every first of January, I organised an ice swim at noon. First it was just me, then me and a few of my friends. Then I started letting people know, through the newspapers, that there would be a hot tub on site. And then suddenly we were not just me and my five friends, we were 25 people, 26 people, 43 people, 70 people, 140 people. When I left in 1994 there were 210 ice bathers.*
> Trygve Bauge

Grandpa Bredo's Tuff Shed. March 2007; **p129**

KrioRus Part 2
Alabushevo, Moscow

Temporary storage container. September 2010; **p133**
During the summer of 2010 KrioRus designed and built their first cryostat, modelled on the ones at the Cryonics Institute. On Wednesday 15 September they transferred their patients from a temporary storage container to their new cryostat.

KrioRus technicians outside the facility. September 2010; **p135**

Margaret Kiseleva, holding a photograph of her mother, Ludmila. September 2010; **p137**
Ludmila was originally cryopreserved on 5 July 2009. Margaret returned to KrioRus with her father, Eugenij, to see her mother being transferred to the newly built cryostat.

Patient #2. September 2010; **p138-139**

Pet dog belonging to Slovenian brothers. September 2010; **p141**

Prayer cards. September 2010; **p143**

Patient #4. September 2010; **p145**

Eugenij Kiselev (left) helps technicians hoist his wife Ludmila's body into the new cryostat. September 2010; **p146-147**
Cryostat inventory: four whole body patients, one neuro patient, two dogs (one is head only), one cat and 40 DNA samples.

Timeship
Comfort, Texas

Signpost for Comfort. December 2012; **p151**

Timeship #1. December 2012; **p152**

Timeship #2. December 2012; **p153**

Timeship #3. December 2012; **p155**

Timeship #4. December 2012; **p157**

Timeship #5. December 2012; **p159**

Timeship #6. December 2012; **p160**

Timeship #7. December 2012; **p161**

Ted's Archive
Phoenix, Arizona

Ted, holding a model of the Saturn 5 rocket, at his home in Phoenix, Arizona. December 2012; **p165**

1&2
East Washington Street Building. 1965; **p166**
Ed Hope self-contracted the 15,000 sq ft building in 1966 to setup the Cryo-Care Equipment Corporation.

3&4
Barbara Kraver, Ed Hope, Frank "Rick" Rickenbacker and Ted Kraver with the Cryo-Care trailer at a truck stop. December 1965; **p166**
Ed pulled it with a big convertible all the way to Philadelphia and it was on the Mike Douglas Show.

5&6
First cryocapsule and supplies on the trailer. December 1965; **p166**
The trailer was fitted out to haul either capsules or a large aluminium tank, surplus from a SAC KC-97 refuelling aircraft. This tank transported bulk liquid nitrogen to our facility to refill our cryocapsules when they got low. The transfer tank only had fibreglass insulation, but the high boil-off was transient and of no concern.

Once, Ed was hauling a load through downtown Phoenix, and had to hit the brakes. The tank tipped on its legs and liquid nitrogen spilled and created a low cloud over the road. Ed hightailed it to the facility before he had to do any explaining.

Another time, during a refuelling at Cryo-Care, the trailer was backed up to the loading dock and a long half-inch copper tube was connected to the capsule. Ed forgot that the trailer was not unhitched from his convertible. He had to go somewhere in a hurry, jumped in his car, hit the gas. The copper line went taut and the capsule started rumbling toward the four-foot loading ramp drop-off. Ted screamed: 'EEEEED!' And he stopped his car with the capsule about half way to taking a tumble.

7-11
Ed and Barbara rolling tank. 1965; **p167**
Ed Hope and Barbara Kraver wrapping 'super vacuum insulation' on the first inner capsule that is mounted on a rotating rig. The fibreglass is being interlaced with the very thin aluminium foil. Ted can barely be seen on the floor under the fibreglass roll on the rig feeding the aluminium foil as the tank turns. He believes that there were about 100 layers of the vacuum insulation.
Ed Hope's primary business was wigs, which is why you see the change in Barbara's appearance.

12-19
p168
In April 1966, Cryo-Care was contacted by a man in the Los Angeles area who wanted his recently deceased mother frozen. She had been embalmed and stored in a mortuary refrigerator for about two months. Cryo-Care, which was never set up to store patients, reluctantly proceeded with the cryopreservation. This was the first human cryonics-related freezing. The woman remained in the capsule for a few months at the Cyro-Care facility, then her son decided to have her removed from preservation and buried.

12-18
Ted opening the first capsule (taken on a Minox spy camera). 1966; **p168**
Ted was responsible for the dreaded 48 nuts that hold on both the inner and outer tank heads. The outer was all steel and had a flat rubber gasket. The inner tank was all aluminium and had a flat lead gasket with 48 holes to accommodate the 48 bolts welded onto the flange of the tank. The nuts were also aluminium to affix the dome head to the tank. Ted spent hours screwing and unscrewing nuts on these tanks with a speed wrench. This is the first cryocapsule, which held the first patient.

19
The first woman frozen being removed for burial. 1966; **p168**
This picture was taken to prove she was in the first cryocapsule.

20-27
Cryopreservation of Dr Bedford. January 1967; **p169**
Dr James Bedford was a University of California psychology professor who wrote several books on occupational counselling. He is considered to be the first 'true' cryopreservation.

In June 1965, the Life Extension Society offered the opportunity to preserve one person free of charge. Bedford took the opportunity and was established as their candidate. He suffered from kidney cancer that had later migrated to his lungs. On 12 January 1967, at the age of 73, he was cryo-preserved a few hours after his death.

Though crude, it differed from the previous unnamed female patient in that it was planned in advance and some attempt was made at perfusing the body at the time of clinical death. Dr Bedford remains cryo-preserved at the Alcor Life Extension Foundation.

20
Dr Bedford received in a coffin that had been packed in dry ice. January 1967; **p169**

21
The first second generation cryocapsule assembled, but prior to insulation. January 1967; **p169**

22
Dr Bedford inner tank all wrapped up, ready for closure, prior to pulling vacuum in outer tank and filling with liquid nitrogen. January 1967; **p169**

23
Unfinished, second generation, cryocapsule with a vacuum pump and liquid nitrogen dewar. January 1967; **p169**

24&25
p169
Dr Bedford on rack ready to be closed up for first liquid nitrogen freezing, using a hybrid capsule that had a bolted inner tank from the first generation and an O-ring sealed end cover of the first second generation outer tank.

26&27
Dr Bedford on rack prior to being inserted in hybrid cryocapsule. January 1967; **p169**

28
First generation cryocapsule, open. March 1968; **p170**
Showing inner capsule with aluminium bolts, superinsulation, dam for LN_2, rails for body rack, liquid nitrogen level sensor, and thermocouple in vent line.

29
Second generation cryocapsule with second model instrumentation panel. March 1968; **p170**

30&31
Second-generation cryocapsule with third model instrumentation panel. 1968; **p170**

32
The original Bedford outer steel tank. 1968; **p170**

33
Working on Bedford tank, Rick (left), Ed (scratching his head) and a technician. 1968; **p170**

34-39
Dr Bedford Transfer. 1968; **p171**
The inner tank on Dr Bedford's original

cryocapsule was leaky and frequent pumping was required to maintain a vacuum. He was transferred to a new stainless steel inner tank, which was welded on, rather than bolted on. The welding was a delicate operation with Bedford inside, but one that, if done correctly, ensured no leaks.

34&35
Second generation Bedford cryocapsule; the insulation is in place. 1968; **p171**
One of Ed Hope's wig shop employees looks on.

36&37
Testing the body rack to ensure that it slides, inside and out, and fits correctly. 1968; **p171**

38&39
Dr Bedford on the first body rack, prior to rewrapping and loading on the second body rack. 1968; **p171**

Cryostats #2, Cryonics Institute, Clinton Township, Michigan, USA. April 2010; **p179**

Robert C. W. Ettinger at his home in Clinton Township, Michigan. April 2010; **p180**
Robert C.W. Ettinger is widely regarded as the father of cryonics. In 1962 he wrote *The Prospect of Immortality*, the book largely credited with launching the cryonics movement. He died on the 23rd July 2011, at the age of 92, and became the Cryonics Institute's 106th patient.

> *The date of possible or probable revival will obviously depend on how rapidly technology develops and by what method and by what circumstances the patients are frozen. In general it will be a case of last in first out and by that I mean that the patients most recently frozen, by more advanced methods, will probably not have to wait as long as those frozen earlier by cruder methods… If we had to make a guess, the usual guesses are that patients frozen already by current or previous methods, under relatively favourable circumstances, will have to wait anywhere from 50 to 150 years. Of course that's only guesswork, but to the patient it doesn't really matter how long he really waits, he's not really conscious and of course most people would rather be revived at an earlier date, than a later date, because they would rather be revived at a time when changes have not been so extreme. That's one of our problems of course, we anticipate that there will be radical changes in society and in humanity over the centuries and of course that scares people. People don't want radical change so that's one of the mistakes I made when I wrote my second book, 'Man into Superman'. I laid out some of the potentialities for future life, future existence, hoping that would inspire some recruits for cryonics but of course, that was a mistake. As I said, people don't want radical change, they want the present, but better, gold plated or chocolate covered. They want the present without some of the discomforts and inconveniences and that of course will probably be available for those that want it… There will not be just one future, but many futures. Many of us will select or create our own futures. Some people of course, for some mysterious reason, think that they should decide whether they should live or die based on my ideas about possible futures, which of course is pretty silly, because I'm not going to build their future for them. They will build their own futures or all of us together will build a variety of futures. In any event the main thing is that if you're alive you have possibilities, if you're dead you have nothing. People talk about fear of death, actually people for practical purposes do not fear death. If there's an immediate or obvious threat then people will react to it, but in modern circumstances the threats are not immediate or obvious and people don't react to them. They just go along with what they think is the inevitable, and just accept their fates. But those who have the imagination to challenge what they have been taught, and prefer life over death, and prefer to experience and explore whatever future they choose to build, those people we offer an opportunity.*
> Robert C.W. Ettinger

Acknowledgements

Firstly, I would like to thank Robert C.W. Ettinger, a pioneer without whom this project would not have been possible.

I would like to thank all the cryonicists. Without your trust and willingness to engage with me, this book simply wouldn't exist. I am deeply grateful to each and every one of you.

In the UK: Mike Carter, Maria Camacho, Tim Gibson, Graham Hipkiss, Alan and the late Silvia Sinclair, David Styles, John and Chrissie de Rivaz. Special thanks to Alan, without your help this project would have never got off the ground.

Alcor Life Extension Foundation: Jennifer Chapman, Diane Cremeens, Aaron Drake, Steve Graber, Hugh Hixon, Bonnie Magee, Max More, Natasha Vita More, Dr Mike Perry, Chana Phaedra, Jerry Searcy and D'Bora Tarrant.

Cryonics Institute: Ben Best, Andy Zawacki and Dennis Kowalski for granting me permission to reproduce Robert Ettinger's micrographs.

KrioRus: Mikhail Batin, Galina Fedorenko, Pawel Ilyin, Margaret Kiseleva, Stas Lipin, Danila Medvedev, Kesha Osadchyy, Valerija Pride and Eugene Zimin.

The team at Suspended Animation, especially: Catherine Baldwin and Matthew Sullivan.

And: Trygve Bauge, John Bull, Fred Coates, Jane Curtis Gazit, Rémy Martinot, Aud Morstoel, Anders Sandberg, Bo Shaffer, Stephen Valentine, George Vogler and Shannon Vyff.

Special thanks to Ted Kraver for being incredibly generous with his time and allowing me to reproduce his photographs.

I started this project when I was a student at the University of Brighton. I'd like to thank my tutors: Jim Cooke, Denis Doran, Fergus Heron, Judith Katz, Xavier Ribas, Aaron Schuman, Christopher Stewart, Esther Teichmann, Julia Winckler and Julia Balme. As well as my peers, particularly those who I exhibited the early pictures with after graduating: Alice Achache, Ben Bailey, Simon Carruthers, Richard Chivers, Alex Currie, Annimeke Goldswain-Hein, Harley Kilburn and Heather Tait.

Thanks to those who processed my film and made prints over the years: Sharon Marysmith at FourCorners; everyone at Spectrum Photographic, particularly Klair Bird, Andy Ford and Paul Lowe; Cos and John at Labyrinth Photographic.

The making of this work was partly funded by magazines and newspapers. Thanks to: Ruth Eichorn and Elisabeth Trautnitz at GEO; Kate Edwards at The Guardian Weekend; Annalee Mather at the The Independent; David Barnett and Steve Peck at Wired UK and Marie-Pierre Subtil at 6 Mois.

Halfway through this project I was fortunate enough to take part in a Redeye workshop hosted by Impressions Gallery. This came at just the right time and injected some much-needed momentum. Thanks to: Roxana Alison, Caitriona Dunnet, Jill Cole, John Cuttriss, Debbie Harman, Paul Herrmann, Seba Kurtis, Chris Nunn, Zuzana Salajkova, Alison Stolwood, Teresa Toms, Helen Warburton, A J Wilkinson and Bella Wood for making it such a fulfilling experience.

Thanks to Nigel, Jane and Amy Thompson for putting me up in Ann Arbor, especially as it was for much longer than planned due to a cloud of volcanic ash preventing me from flying home.

I'm sincerely grateful to those artists, curators and editors that showed an interest in the project and offered their advice, in roughly chronological order: Simon Norfolk; Lauren Heinz; Welmer Keesmaat; Brett Rodgers, Clare Grafik and Karen McQuaid at The Photographers' Gallery; Andy Adams; Jake Stangel; Michael Mack; Alec Soth; Pete Brook; Alistair Robinson at the Northern Gallery for Contemporary Art; Sophie Balhetchet and Gregory Barker at HotShoe; Martin Parr; Sarah Cook and Judith Alder.

Thanks to Impressions Gallery: Pippa Oldfield, Sarah Read, Angela Sheard, Jennifer Sobol and especially Anne McNeill who saw potential in the project and worked with me to transform it into an exhibition.

I feel incredibly fortunate to have worked with such brilliant and knowledgeable people on my first book. Thanks to everyone at GOST books: Allon Kaye, Gordon MacDonald, Eleanor McNair, Claudia Paladini, Justine Schuster and Stuart Smith - who kept this book alive when I had come close to giving up.

Thanks to my dear friends and supporters, particularly: Jonas Bendiksen, Jo Coles, Jason Grima, Nina Emett, Kathryn Hall, Harriet Harmer, Liz Hingley, Sarah Lloyd, Lydia and Kieran O'Callaghan, Oliver Perrott, Simon Roberts, Jim Stephenson and the White family. And a very special thanks to my lovely family: Mum, Dad, Jade, Cameron and most of all Gabriella for her love, support and never holding back from offering her honest criticism.

Finally, I would like to dedicate this book to Mark Power, for being there from the very beginning and offering his valuable guidance throughout. I'm eternally grateful for everything you have taught me about life and photography.

Murray Ballard
Brighton, February 2016

First published in 2016 by
GOST Books
8a West Smithfield
London EC1A 9JR
info@gostbooks.com
gostbooks.com

Edited and designed by GOST
and Murray Ballard
Printed in Italy by EBS

British Library cataloguing-in-publication data. A catalogue record of this book is available from the British Library.

ISBN 978-1-910401-03-3

A GOST Books Production

GOST

The tired old man, then, will close his eyes, and he can think of his impending temporary death as another period under anaesthesia in the hospital. Centuries may pass, but to him there will be only a moment of sleep without dreams.

Robert C.W. Ettinger

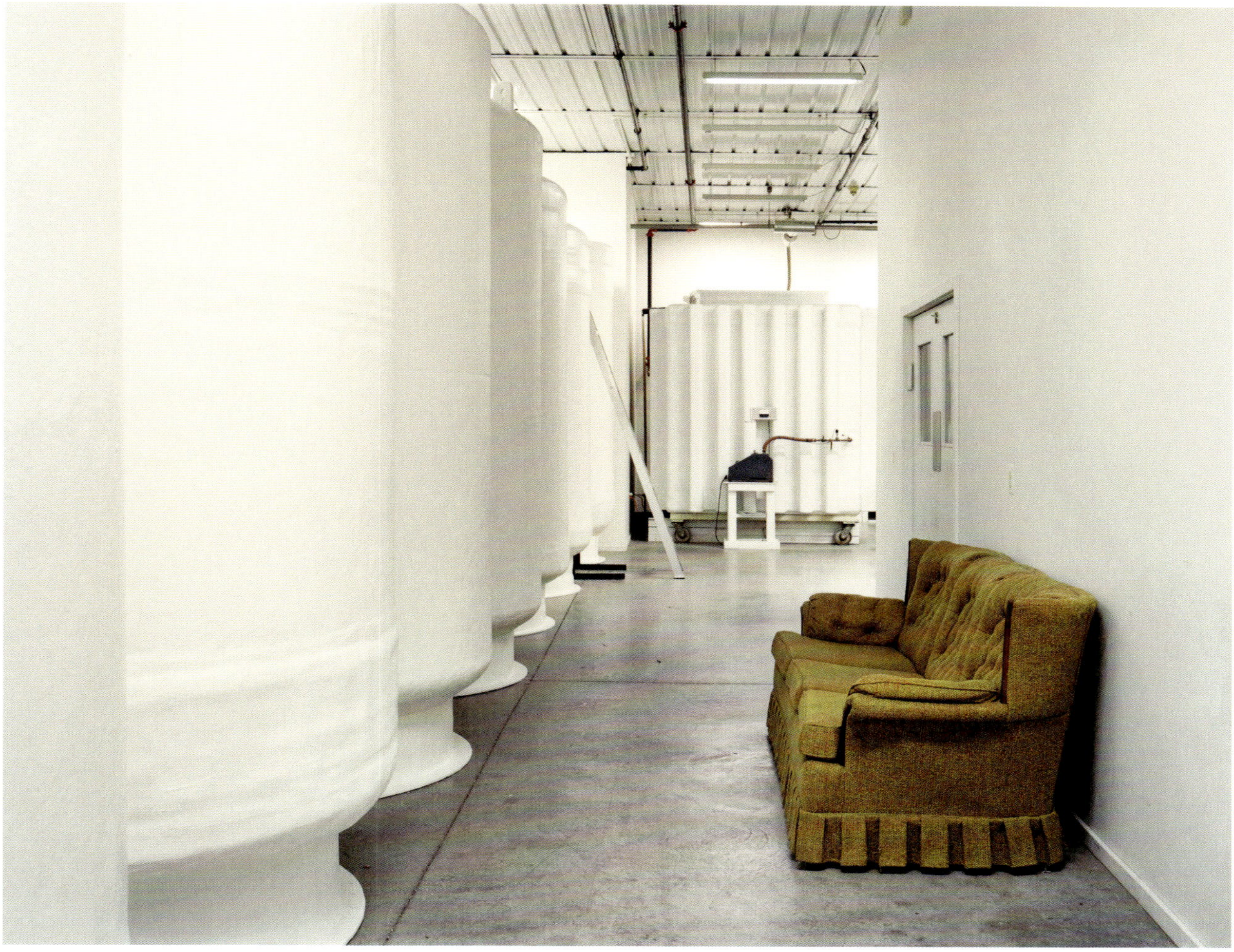

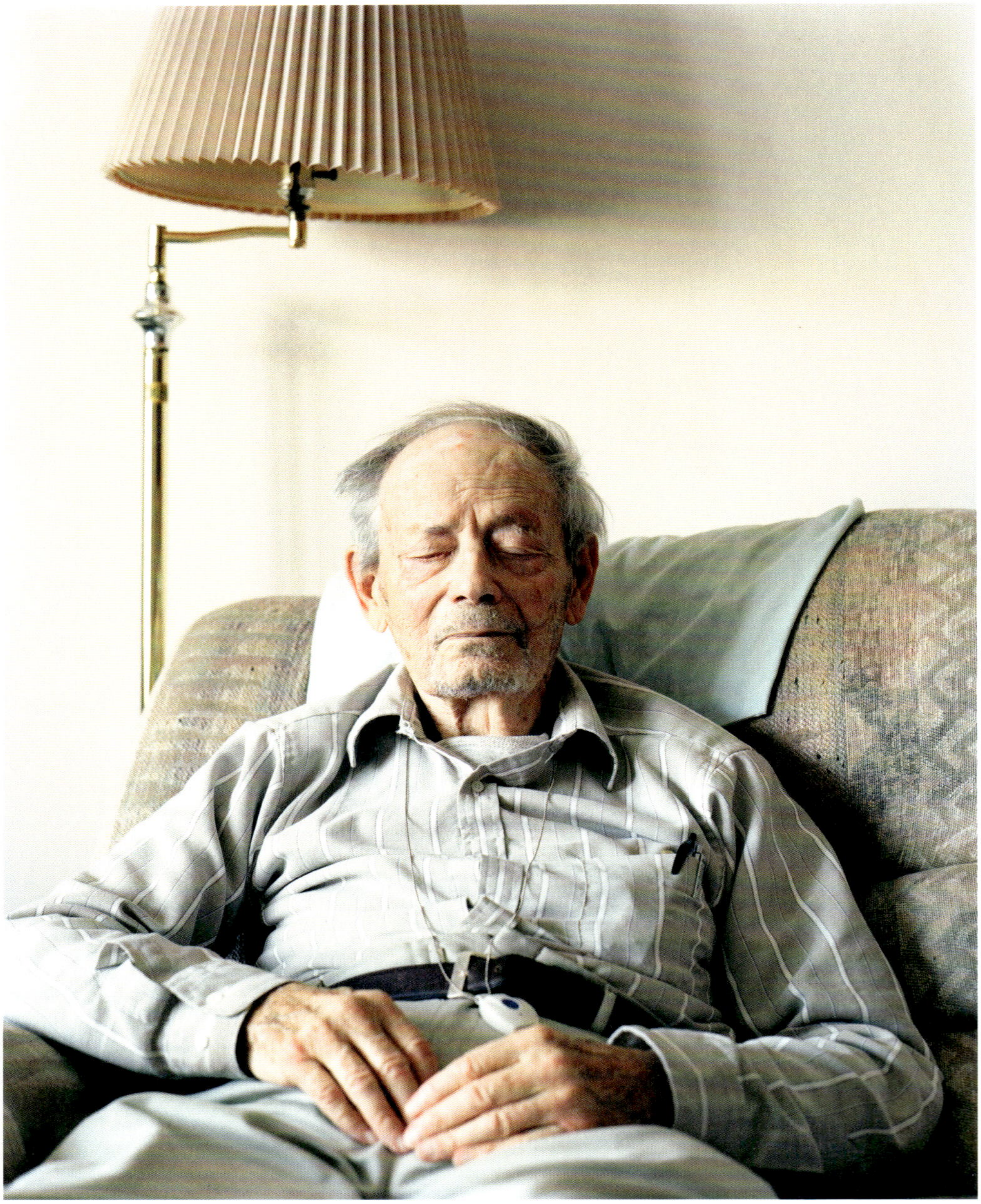